NUCLEAR–RENEWABLE
HYBRID ENERGY SYSTEMS

The following States are Members of the International Atomic Energy Agency:

AFGHANISTAN
ALBANIA
ALGERIA
ANGOLA
ANTIGUA AND BARBUDA
ARGENTINA
ARMENIA
AUSTRALIA
AUSTRIA
AZERBAIJAN
BAHAMAS
BAHRAIN
BANGLADESH
BARBADOS
BELARUS
BELGIUM
BELIZE
BENIN
BOLIVIA, PLURINATIONAL
   STATE OF
BOSNIA AND HERZEGOVINA
BOTSWANA
BRAZIL
BRUNEI DARUSSALAM
BULGARIA
BURKINA FASO
BURUNDI
CAMBODIA
CAMEROON
CANADA
CENTRAL AFRICAN
   REPUBLIC
CHAD
CHILE
CHINA
COLOMBIA
COMOROS
CONGO
COSTA RICA
CÔTE D'IVOIRE
CROATIA
CUBA
CYPRUS
CZECH REPUBLIC
DEMOCRATIC REPUBLIC
   OF THE CONGO
DENMARK
DJIBOUTI
DOMINICA
DOMINICAN REPUBLIC
ECUADOR
EGYPT
EL SALVADOR
ERITREA
ESTONIA
ESWATINI
ETHIOPIA
FIJI
FINLAND
FRANCE
GABON
GEORGIA

GERMANY
GHANA
GREECE
GRENADA
GUATEMALA
GUYANA
HAITI
HOLY SEE
HONDURAS
HUNGARY
ICELAND
INDIA
INDONESIA
IRAN, ISLAMIC REPUBLIC OF
IRAQ
IRELAND
ISRAEL
ITALY
JAMAICA
JAPAN
JORDAN
KAZAKHSTAN
KENYA
KOREA, REPUBLIC OF
KUWAIT
KYRGYZSTAN
LAO PEOPLE'S DEMOCRATIC
   REPUBLIC
LATVIA
LEBANON
LESOTHO
LIBERIA
LIBYA
LIECHTENSTEIN
LITHUANIA
LUXEMBOURG
MADAGASCAR
MALAWI
MALAYSIA
MALI
MALTA
MARSHALL ISLANDS
MAURITANIA
MAURITIUS
MEXICO
MONACO
MONGOLIA
MONTENEGRO
MOROCCO
MOZAMBIQUE
MYANMAR
NAMIBIA
NEPAL
NETHERLANDS
NEW ZEALAND
NICARAGUA
NIGER
NIGERIA
NORTH MACEDONIA
NORWAY
OMAN
PAKISTAN

PALAU
PANAMA
PAPUA NEW GUINEA
PARAGUAY
PERU
PHILIPPINES
POLAND
PORTUGAL
QATAR
REPUBLIC OF MOLDOVA
ROMANIA
RUSSIAN FEDERATION
RWANDA
SAINT KITTS AND NEVIS
SAINT LUCIA
SAINT VINCENT AND
   THE GRENADINES
SAMOA
SAN MARINO
SAUDI ARABIA
SENEGAL
SERBIA
SEYCHELLES
SIERRA LEONE
SINGAPORE
SLOVAKIA
SLOVENIA
SOUTH AFRICA
SPAIN
SRI LANKA
SUDAN
SWEDEN
SWITZERLAND
SYRIAN ARAB REPUBLIC
TAJIKISTAN
THAILAND
TOGO
TONGA
TRINIDAD AND TOBAGO
TUNISIA
TÜRKİYE
TURKMENISTAN
UGANDA
UKRAINE
UNITED ARAB EMIRATES
UNITED KINGDOM OF
   GREAT BRITAIN AND
   NORTHERN IRELAND
UNITED REPUBLIC
   OF TANZANIA
UNITED STATES OF AMERICA
URUGUAY
UZBEKISTAN
VANUATU
VENEZUELA, BOLIVARIAN
   REPUBLIC OF
VIET NAM
YEMEN
ZAMBIA
ZIMBABWE

The Agency's Statute was approved on 23 October 1956 by the Conference on the Statute of the IAEA held at United Nations Headquarters, New York; it entered into force on 29 July 1957. The Headquarters of the Agency are situated in Vienna. Its principal objective is "to accelerate and enlarge the contribution of atomic energy to peace, health and prosperity throughout the world".

IAEA NUCLEAR ENERGY SERIES NR-T-1.24

NUCLEAR–RENEWABLE HYBRID ENERGY SYSTEMS

INTERNATIONAL ATOMIC ENERGY AGENCY
VIENNA, 2022

COPYRIGHT NOTICE

All IAEA scientific and technical publications are protected by the terms of the Universal Copyright Convention as adopted in 1952 (Berne) and as revised in 1972 (Paris). The copyright has since been extended by the World Intellectual Property Organization (Geneva) to include electronic and virtual intellectual property. Permission to use whole or parts of texts contained in IAEA publications in printed or electronic form must be obtained and is usually subject to royalty agreements. Proposals for non-commercial reproductions and translations are welcomed and considered on a case-by-case basis. Enquiries should be addressed to the IAEA Publishing Section at:

Marketing and Sales Unit, Publishing Section
International Atomic Energy Agency
Vienna International Centre
PO Box 100
1400 Vienna, Austria
fax: +43 1 26007 22529
tel.: +43 1 2600 22417
email: sales.publications@iaea.org
www.iaea.org/publications

© IAEA, 2022

Printed by the IAEA in Austria
December 2022
STI/PUB/2041

IAEA Library Cataloguing in Publication Data

Names: International Atomic Energy Agency.
Title: Nuclear–renewable hybrid energy systems / International Atomic Energy Agency.
Description: Vienna : International Atomic Energy Agency, 2022. | Series: nuclear energy series, ISSN 1995–7807 ; no. NR-T-1.24 | Includes bibliographical references.
Identifiers: IAEAL 22-01563 | ISBN 978–92–0–148922–7 (paperback : alk. paper) | ISBN 978–92–0–149022–3 (pdf) | ISBN 978–92–0–149122–0 (epub)
Subjects: LCSH: Renewable energy sources. | Hybrid power systems. | Nuclear energy. | Energy conservation.
Classification: UDC 621.039:502.174.3 | STI/PUB/2041

FOREWORD

The IAEA's statutory role is to "seek to accelerate and enlarge the contribution of atomic energy to peace, health and prosperity throughout the world". Among other functions, the IAEA is authorized to "foster the exchange of scientific and technical information on peaceful uses of atomic energy". One way this is achieved is through a range of technical publications including the IAEA Nuclear Energy Series.

The IAEA Nuclear Energy Series comprises publications designed to further the use of nuclear technologies in support of sustainable development, to advance nuclear science and technology, catalyse innovation and build capacity to support the existing and expanded use of nuclear power and nuclear science applications. The publications include information covering all policy, technological and management aspects of the definition and implementation of activities involving the peaceful use of nuclear technology. While the guidance provided in IAEA Nuclear Energy Series publications does not constitute Member States' consensus, it has undergone internal peer review and been made available to Member States for comment prior to publication.

The IAEA safety standards establish fundamental principles, requirements and recommendations to ensure nuclear safety and serve as a global reference for protecting people and the environment from harmful effects of ionizing radiation.

When IAEA Nuclear Energy Series publications address safety, it is ensured that the IAEA safety standards are referred to as the current boundary conditions for the application of nuclear technology.

Two principal options for low carbon energy are renewables and nuclear energy. While many Members States have expressed interest in these options, possible synergies between them and potential integration options have not been fully explored. Nuclear–renewable hybrid energy systems integrate these energy generation sources to leverage the benefits of each technology for improved reliability and sustainability. Nuclear–renewable hybrid energy systems can produce heat, electricity and other products that society requires while supporting higher penetrations of variable renewable generation (i.e. wind and solar photovoltaics). Nuclear–renewable hybrid energy systems can include various applications, such as seawater desalination, hydrogen production, district heating or cooling, the extraction of tertiary oil resources and process heat applications, such as cogeneration, coal to liquid conversion and assistance in the synthesis of chemical feedstock.

In October 2018, the IAEA held a Technical Meeting on Nuclear–Renewable Hybrid Energy Systems for Decarbonized Energy Production and Cogeneration. This meeting proposed the development of an IAEA publication on nuclear–renewable hybrid energy systems. This publication provides high level information to decision makers and stakeholders, including that necessary when considering nuclear–renewable hybrid energy systems.

This publication presents opportunities for nuclear–renewable hybrid energy systems that could be pursued in various Member States as a part of their future energy mix. It describes the motivation for and potential benefits of nuclear–renewable hybrid energy systems relative to independent nuclear and renewable generation producing electricity alone. Considerations for implementation are outlined in the publication, including gaps that require additional technology and regulatory development. This publication intends to equip decision makers and stakeholders with sufficient information to consider nuclear–renewable hybrid energy systems as an option within regional and national energy systems.

The IAEA officers responsible for this publication were T. Jevremovic of the Division of Nuclear Power and A. van Heek of the Division of Planning, Information and Knowledge Management.

CONTENTS

SUMMARY

Nuclear energy and renewables are the two principal options for low carbon energy generation. However, synergies among these resources have yet to be fully exploited, and the advantages of integrating these generation options directly are only now being explored. Nuclear–renewable hybrid energy systems (HESs) consider opportunities to couple these energy generation sources to leverage the benefits of each technology to provide reliable, sustainable electricity to the grid and to provide low carbon energy to other energy use sectors.

The transition of the global energy mix to include increasing fractions of variable renewable energy resources is driven by economics as well as social development concerns. While introducing new challenges, this transition also presents potential synergies and opportunities for sustainable development. In particular, the proposed coupling and/or tighter integration of nuclear and renewable resources appear to be mutually beneficial.

This publication describes the potential use of nuclear and renewable generation in coordinated, and in some cases tightly coupled, configurations to support various applications beyond electricity production, including seawater or brackish water desalination, hydrogen production, district heating or cooling, the extraction of tertiary oil resources, and process heat applications such as cogeneration, coal to liquids conversion and assistance in the synthesis of chemical feedstock. Where available, case studies are presented for these configurations to describe relevant market conditions and trends, energy requirements and research gaps in order to clarify the opportunities and issues associated with the proposed nuclear–renewable HESs.

Considerations for implementation are outlined, including gaps that require additional technology and regulatory developments. This publication intends to equip decision makers and stakeholders with sufficient information to consider nuclear–renewable HESs as an option within regional and national energy systems. While the true value of these multi-input, multioutput energy systems remains to be demonstrated, and may differ as a function of deployment region and energy market structures, research to date suggests that nuclear–renewable HES may play a key role in meeting future energy demands in a manner that provides flexibility and resilience, while supporting established sustainable development goals.

1. INTRODUCTION

1.1. BACKGROUND

Two principal options for low carbon energy are renewables and nuclear energy. The synergies between them and the advantages of integrating these options have not yet been fully explored. Nuclear–renewable HESs take advantage of coupling these energy generation sources to leverage the benefits of each technology to provide reliable, sustainable electricity to the grid and also provide low carbon energy to other energy use sectors. Nuclear–renewable HESs can produce heat, electricity and other products that society requires while supporting higher penetrations of variable renewable generation. In this manner, nuclear–renewable HESs can provide energy to support various applications, such as seawater or brackish water desalination, hydrogen production, district heating or cooling, the extraction of tertiary oil resources, and process heat applications such as cogeneration, coal to liquids conversion and assistance in the synthesis of chemical feedstock.

Nuclear–renewable HESs are defined as integrated facilities comprising nuclear reactors, renewable energy generation and industrial processes that can simultaneously address the need for grid flexibility, greenhouse gas (GHG) emission reductions and optimal use of investment capital. These systems are often referred to as integrated energy systems that incorporate multiple generators and produce multiple energy products in either coordinated systems (i.e. a loosely coupled network of generators in a grid balancing area) or tightly coupled energy systems (i.e. an energy park scenario in which subsystems are codesigned and cocontrolled). Current energy systems are primarily loosely coupled via grid interactions, having a primary focus on supporting electricity demand. The adoption of tightly coupled designs that incorporate multiple thermal and electrical generators to support electric and non-electric energy demands offers an opportunity to optimize energy use in real time to better utilize energy generation assets, thereby using invested capital more efficiently. Focusing on the use of low emission generation resources in these highly efficient energy parks provides the further benefit of reducing emissions across multiple energy use sectors. Figure 1 illustrates the basic structure of a nuclear–renewable HES comprising nuclear reactors, renewables, industrial processes and grid interconnection (Annex I provides the legend for the graphics used in this publication).

1.2. OBJECTIVE

This publication presents opportunities for nuclear–renewable HESs that are or could be pursued in various Member States as part of their present and future energy mix. It describes the motivation for and

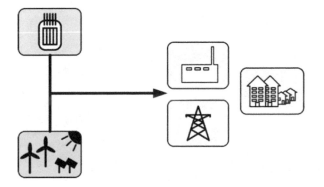

FIG. 1. Generic representation of a nuclear–renewable HES.

potential benefits of nuclear–renewable HESs relative to independent nuclear and renewable generation producing electricity alone. Considerations for implementation are outlined, including gaps that require additional technology and regulatory developments. This publication intends to equip decision makers and stakeholders with sufficient information to consider nuclear–renewable HESs as an option within regional and national energy systems.

1.3. SCOPE

This publication provides information on nuclear–renewable HESs for stakeholders in academia, industry, government agencies and public institutions. The guidance provided here, describing good practices, represents expert opinion but does not constitute recommendations made on the basis of a consensus of Member States. An additional publication will be made available summarizing the Technical Meeting and discussions at a technical level.

1.4. STRUCTURE

Section 2 reviews the present energy market. How this energy market may profit from the near term deployment of nuclear–renewable HESs is investigated in Section 3, while Section 4 provides definitions for nuclear–renewable HESs that are extended with case studies in Section 5. The case studies are followed by a discussion of considerations for implementation (Section 6), anticipated challenges and technology gaps (Section 7), and conclusions and recommendations (Section 8). Annex I provides the legend for the graphics used in this publication.

1.5. USERS

The primary users of this publication are utilities/operator organizations, governing organizations, or others who are or will be responsible for national nuclear power programme development.

Technical experts of Member States, regulators, and others actively involved in planning and developing a nuclear power programme and nuclear power project, and technical experts involved in advising government or utility officials ought to benefit from this publication regarding impacts on planning and national infrastructure development.

2. STATUS QUO IN ENERGY MARKETS: TRENDS AND EVOLUTION

2.1. LOW CARBON ENERGY FUTURE

Since 2000, global primary energy demand has increased by more than 40%, amounting to 14 314 Mtoe[1] (~166.5 PWh) in 2018. Some energy scenarios are predicting that increasing incomes and 1.7 billion additional people will push up demand by almost 25% by 2040 [1]. The growth is expected to be mainly driven by developing countries as people such as the 500 million in sub-Saharan Africa gain access to electricity [2]. Ensuring universal access to affordable, reliable and modern energy services is one of the key Sustainable Development Goals [2]. Energy is essential for the improvement of living standards and wellbeing, lifting people out of poverty, creating new jobs, ensuring food security, and improving societal cohesion and inclusiveness.

At the same time, the United Nations (UN) Intergovernmental Panel on Climate Change (IPCC) has estimated that human activities have led to approximately 1°C of global warming above pre-industrial levels. The intensity and frequency of some climate and weather extremes, such as sea level rise and hot weather conditions, have increased [3].

The IPCC study concludes, with high confidence, that it is imperative to keep the global temperature rise to less than 1.5°C above pre-industrial levels to minimize climate related risks to health, livelihood, human security, food security, water supply and economic growth. These issues are expected to be even more pervasive if the global temperature increase reaches 2°C. Limiting the temperature increase requires limiting of anthropogenic greenhouse gas emissions, especially in the energy sector. Mitigation strategies involve lowering energy and resource intensity, including enhanced energy efficiency, and deployment of low emission energy sources, such as renewables and nuclear. In the long term perspective, carbon dioxide (CO_2) removal strategies are also considered.

2.2. NUCLEAR ENERGY GROWTH TRENDS

In 2019, slightly over 10% of the world's electricity, 2586 TWh, was generated by approximately 450 nuclear power plants (NPPs) operating in 30 countries worldwide. Their total net capacity was of ~400 GW(e) [4]. Fifteen countries are currently constructing new nuclear power reactors. At the end of 2019, 57 GWe NPP capacity was under construction, comprising 54 reactors. The largest number of these reactors are being built in China, India, the Russian Federation and the United Arab Emirates [5]. Nuclear energy is currently used almost exclusively to produce electricity.

The 2019 edition of the International Energy Agency (IEA) World Energy Outlook (WEO) [1] does not provide a forecast of the future energy system. Rather it includes three scenarios:

(a) The Current Policies Scenario that estimates a likely future based on existing policies;
(b) The Stated Policies Scenario that incorporates policy plans and announcements into the estimates; and
(c) The Sustainable Development Scenario that identifies a way to meet emission reduction goals.

The Stated Policies Scenario estimates installed nuclear capacity growth of over 15% from 2018 to 2040 (reaching ~482 GW(e)). The scenario envisages a total generating capacity of 13 109 GW(e) by 2040, with

---

[1] The tonne of oil equivalent (toe) is a unit of energy defined as the amount of energy released by burning one tonne of crude oil; it is approximately 42 gigajoules (GJ) or 11.63 megawatt-hours (MWh).

the increase heavily concentrated in Asia, and in particular China (34% of the total). In this scenario, nuclear's contribution to global power generation grows from 2718 TWh to 3475 TWh, an increase of 28% ,which is greater than the capacity growth due to projected capacity factor increases. Nuclear's projected generation share is approximately 8.5% in 2040. Although total demand and power generation are increasing, the relative contribution of nuclear is decreasing slightly as a result of the growth of other energy generation resources.

2.3. RENEWABLE ENERGY GROWTH TRENDS

According to the 2019 edition of the IEA WEO, renewable energy will experience the largest growth rates of all energy generation technologies in the coming decades. In the Stated Policies Scenario total primary renewable utilization is projected to grow by 125% between 2018 and 2040, from 1391 Mtoe to 3127 Mtoe [1]. In that same scenario, renewable electricity generation (from hydropower, bioenergy, wind, geothermal, solar and marine technologies) will grow by 165% from 6799 TWh in 2018 to 18 049 TWh in 2040. Wind and solar photovoltaic (PV) systems are both projected to grow by over 4000 TWh/year, from 5226 TWh/year in 2018 to 4705 TWh/year in 2040.

The capacity of solar PV is projected to grow from 495 GW to 3142, accounting for 24% of all new generation capacity added by 2040. Energy generation from wind turbines is expected to account for approximately 12% of new generation capacity added by 2040. Both solar PV and wind only produce electricity when the resource is available, resulting in lower capacity factors and limitations on when the resources generate electricity; issues that are important to consider when planning future energy mix.

2.4. ENERGY MIX CONSIDERATIONS

Cost reductions in renewable energy technologies, aided in part by incentives such as feed-in tariffs for renewable technologies, and advances in digital technologies are opening up great opportunities for energy transitions, while creating some new energy security challenges. Wind and solar PV systems are predicted to provide more than half of the additional electricity generation in 2040 in the Stated Policies Scenario of the WEO 2019 and almost all the growth in the Sustainable Development Scenario [4]. Policy makers and regulators will have to move fast to keep up with the pace of technological change and the rising need for flexible operation of power systems that is introduced when a grid has a larger fraction of its capacity supported by variable, non-dispatchable resources.

Nuclear energy offers flexible, low GHG emitting power generation and is not limited to regions with inexpensive fuel supply, as nuclear fuel only accounts for a fraction of a power plant's operational costs. As countries commit to opt out of fossil fuel power generation in favour of low emission options, it is possible that future power generation grids could mostly consist of renewable and nuclear sources. Due to the variable nature of renewable power generation sources, NPPs and energy storage options are likely to play an ever more important role in balancing energy grids.

In addition, primary energy consumed for electricity generation was only 38% of the total global primary energy consumption in 2018 [4] and that percentage is only expected to grow to 40% in the Stated Policies Scenario 2040 projection. Electricity comprises the largest demand for primary energy but industry, transportation, and building and water heating and cooling also involve large demands.

To address the majority of energy and emission issues, policy makers will need to look beyond electricity production and consider options to increase nuclear and renewable options for those other applications.

In addition, policy makers need to consider how to utilize capital investment for energy generation better. In other words, a higher capacity factor for generating energy is more economically efficient than ramping down to provide flexibility. This impact is especially noticeable for high capital, low operating cost options such as nuclear and renewable generation.

3. MOTIVATION FOR A NEW PARADIGM OF NUCLEAR–RENEWABLE HYBRID ENERGY SYSTEMS

3.1. KEY MOTIVATION FOR THE HYBRID ENERGY SYSTEM AS A WHOLE

The transition of the global energy mix to include increasing fractions of variable renewable energy sources (RESs) is driven by economics as well as social development concerns; hence, this transition receives a high level of political support and funding in many countries. While introducing new challenges, this transition also presents potential synergies and opportunities for sustainable development. In particular, the proposed coupling and/or tighter integration of nuclear and renewable resources appears to be mutually beneficial. As discussed in more detail below, the variable nature of renewables can be well compensated by coupling these technologies with nuclear energy to provide a more equitable and inclusive energy system that is also environmentally sustainable, safe, reliable and affordable, while supporting enhanced grid resilience.

3.2. CONSIDERATIONS RELATED TO NUCLEAR ENERGY IN CURRENT MARKETS

3.2.1. Flexible operation of nuclear power plants

NPPs have traditionally been dispatched as baseload generation systems, typically operating at their rated capacity with a load factor of 80–95%, where this load factor is only limited by refuelling and maintenance requirements [6]. This operating mode primarily results from low operating costs and overcomes the usually high capital costs, thereby achieving competitive economics compared to other forms of electricity generation [7, 8].

Large scale addition of variable and decentralized renewable generation (mainly wind and solar photovoltaic) capacities is impacting on electricity market conditions. Historically, electricity supply and demand have typically been balanced on the supply side through reduced load factors of conventional (fossil fuelled and hydroelectric) power plants and increased ramping of their output. Where the installed NPP capacity is substantial in comparison to the minimum energy load of the system and combined with large renewable energy generation, the NPP load would also usually be reduced to balance electricity supply and demand. Specifically, in liberalized wholesale markets using the pay as clear bidding process, the order in which available sources are brought online to meet demand is determined on the basis of the variable costs of the power generation. As wind and solar PV systems have virtually no variable costs, they are typically utilized by the grid whenever they generate, displacing conventional technologies with higher variable generation costs.[2] This mode of operations results in very low and, at times, negative grid electricity prices. The latter is a consequence of a temporary oversupply of electricity, which may occur in particular in off peak hours and in energy systems with high penetration of variable renewable sources. Most operating reactors have a technical capability to follow the load [9].

In some countries, for example in France and Germany, load following is already practised regularly, providing standard grid operator services (i.e. frequency modulation and regulation of the demand variability) as well as flexibility to integrate variable renewable output. Flexible operation of NPP power output to the grid is also becoming more prevalent in regions of the United States of America due to the

---

[2] In many markets, there are also mandatory national targets for the introduction and use of renewable energies in the energy mix.

low cost of natural gas and increased renewable electricity generation that periodically drives electricity prices to become very low or negative (see also Subsection 3.2.2) [10–12].

Many countries also require any new NPP to provide such increased flexibility in relatively fast power variations as well. As an example, in accordance with the European Utility Requirements, any new NPP is expected to be able to operate continuously between its minimum regulating level (~50%) and 100% of its rated output (Pr), scheduled and unscheduled, with a rate of change in its nominal output of 3% Pr/min [13]. Higher manoeuvrability exists for most modern plant designs, with wider power range variations and ramps of up to 5% Pr/min available. Advanced reactor designs in development, which generally incorporate non-water coolants and enhanced safety features, are expected to match or exceed these capabilities. Some nations also consider water cooled reactors that incorporate advanced technologies, such as small modular reactors (SMRs) and enhanced large scale designs, in the advanced reactor category.

3.2.2. Competitive pressure from low cost natural gas

Globally, fossil fuels make the biggest contribution to electricity generation, with the largest contribution being derived from coal fired power plants and the second largest from natural gas [1]. The development of shale gas resources through the combined application of horizontal drilling and hydraulic fracturing (often referred to as fracking) has led to a substantial increase in the world's proven reserves of natural gas and its production and supply on the world market. Since 2000, the world's proven reserves of natural gas have increased by approximately 26%, mainly due to the development of horizontal fracking technology [14]. The world demand for natural gas has grown by more than 50% since 2000 (3273 Mtoe (~38.1 PWh) in 2018) [4]. At the same time, in some countries and regions, gas wholesale prices have reached their lowest point in 20 years [15].

Combined cycle gas turbine (CCGT) plants typically have low fixed operation and maintenance costs, low capital investment needs (<1000 US $/kW [7]), feature short deployment times and are highly flexible, with a power ramping of 20−50% Pr/min possible. Advanced CCGT plants also feature high thermodynamic efficiencies (50−65%) [16]. Thus, with low natural gas prices, electricity can be generated flexibly at low cost. The abundant supply of natural gas and the resulting low cost electricity pushes wholesale power prices down, decreasing current and forecast revenues available from generated electricity. In some cases, revenues have now declined to the level that they are below the operating costs of NPPs [17].

3.3. CONSIDERATIONS FOR INCREASING PENETRATION OF RENEWABLES

In response to commitments made in the 2015 Paris Agreement and/or individual Member State energy policies to reduce GHG emissions, many countries are planning to substantially increase the share of electricity generated by RESs [18]. Considering announced countries policies and targets, it is estimated that the share of global electricity generated by renewables (including non-variable) will increase from 25% in 2018 to approximately 40% by 2040, limiting the corresponding increase in CO_2 emissions to 7% compared to 2018 [4]. However, to put the world on track to meet climate goals and stay well below a 2°C (i.e. 1.7−1.8°C) increase in global mean temperatures from pre-industrial levels, the 2019 IEA WEO estimates that the share of world electricity generated by renewables (including non-variable) would need to further increase to approximately 67% by 2040. This, together with significantly improved energy efficiencies to curb a growth in primary energy demand and combined with increased transmission capacities and energy storage, would deliver a global decrease of energy related CO_2 emissions of approximately 30% by 2040 compared to 2000 [4].

In a much more ambitious regional context, the countries of the European Union (EU) have committed to achieving climate neutrality and a reduction of greenhouse gas emissions of at least 80% in 2050 compared to 1990, with more than 80% of the electricity derived from renewables [19, 20].

These approaches to emission reduction focus on the electricity sector, whereas greater impact on global emissions can be achieved by also considering reducing emission from the industrial and transportation sectors. Considering the limited opportunities to expand the capacity of non-variable RESs because of geographical limitations (e.g. hydroelectric, geothermal, biomass), the majority of newly installed renewable capacity is expected to be variable in nature (mostly wind and solar PV) [4]. The large penetration of variable renewables may, however, lead to increased system cost[3], as reported in the 2019 Nuclear Energy Agency publication [21], reducing the investment value. Significant increases in renewable penetration may also have relatively weak public support in impacted regions (in particular for land based wind power), may lead to reduced energy use efficiencies that result from curtailment at times of overproduction, or transmission of excess power into the transmission grids of neighbouring countries, which can stress those grids [22, 23]. Large fractions of variable RESs in the energy mix typically necessitate additional measures to match supply and demand on the electrical grid due to reduced amounts of dispatchable and flexible generation capacities that can balance them, as has been done historically. These extra measures can involve significant investments in grid infrastructure upgrades, energy storage systems and ancillary services, including improved scheduling and dispatch, frequency and voltage control, and operation of power/spinning reserves.

3.4. DRIVERS AND OPPORTUNITIES

The evolving grid dynamics that arise from increasing variable renewable capacity provide an opportunity for system optimization via integration of nuclear and renewable resources. Such integration may include nuclear cogeneration [24] and implementation of a tight coupling between nuclear and renewables (i.e. via nuclear–renewable HESs). Nuclear cogeneration enables switching between electricity and heat generation, with NPPs performing load following while still operating the reactor at a high capacity factor. At the same time, less curtailment is necessary, with the surplus of energy generated by nuclear–renewable HESs being used to deliver new low carbon energy products (e.g. hydrogen, hydrocarbons, industrial process heat) or new services (e.g. potable water).

3.4.1. Financial drivers

As revenue is generated through the sale of electricity, high quality heat, and/or other energy products and services, capital investment in energy generating sources has the potential to be more profitable and attractive for investors. The generated electricity has the potential to remain affordable and the newly developed products and services have the potential to open up new market opportunities, particularly for industrial and transport applications. These opportunities are stimulated and developed by increasing urgency to decarbonize rapidly, not only in the power sector, but also in the transport and industrial sectors. Additionally, increasing the share of renewables in the energy mix allows for more favourable exploitation of synergies from tighter integration of nuclear and renewable sources.

3.4.2. Environmental drivers

In 2017, approximately 40% of total global energy related CO_2 emissions were due to power production, while the transport and industrial sectors accounted for approximately 25% and 19%, respectively [25]. In the period of 1970–2015, nuclear power avoided approximately 68 Gt of CO_2 emissions, while hydropower and other renewables avoided approximately 90 Gt and 10 Gt of CO_2 emissions, respectively. Apart from electricity and heat, options for reducing CO_2 emissions and thus

---

[3] Based on the OECD/NEA, Nuclear Energy and Renewables: System Effects in Low-carbon Electricity Systems, Technical Report NEA No. 7056, Paris (2012): system costs are defined as the total costs above plant level costs to supply electricity at a given load and given level of security of supply.

limiting the increase of global mean temperatures are limited and require rapid and sustained energy transition in these other contributing sectors [3]. The deployment of nuclear–renewable HESs moves low carbon generation beyond electricity and thus could contribute substantially to the decarbonization of transport and industrial sectors by producing low carbon fuels, feedstock and heat for industry. This would help mitigate the effects of human induced global warming while sustaining further population, economic and societal growth.

Air pollution is an important health and environmental challenge, with the three major air pollutants being sulphur dioxide, nitrogen oxides and fine particulate matter ($PM_{2.5}$). Energy related outdoor air pollution from using fossil fuels caused an estimated three million premature deaths in 2018 [4]. The health impacts of indoor air pollution from using combustible fuels for household energy are even more severe, and were estimated to have caused 4.3 million deaths in 2012; almost all in low and middle income countries [2, 26]. In one estimate for the period 1971–2009 nuclear power prevented a mean value of 1.84 million air pollution related deaths that would have otherwise resulted from fossil fuel combustion [27]. Likewise, nuclear–renewable HESs would produce fewer major air pollutants than options requiring combustion, reducing health impacts such as premature deaths resulting from high air pollution.

3.4.3. Social development drivers

The questions of social and economic development, availability of energy and access to safe potable water are fundamentally interwoven. There are close to one billion people without access to electricity today, while more than a quarter of the world population (2.1 billion people) do not have access to safe drinking water [2, 4]. Both issues are most pronounced in sub-Saharan Africa. Nuclear–renewable HESs could provide low carbon electricity and potable water through desalination in a reliable and affordable manner, enabling economic growth and social development, and fostering domestic jobs, helping to achieve multiple UN Sustainable Development Goals.

3.4.4. Geostrategic drivers

The transition to a low carbon economy is expected to result in an energy system in which primary energy supply is largely derived from renewable energy sources, combined with nuclear power and with a diminishing share of fossil fuels [4]. In delivering low carbon electricity and new energy products, nuclear–renewable HESs can support this transition by displacing fossil fuels for electricity generation and fossil fuel combustion to support the transport and industrial sectors. This would significantly improve security of energy supply, as the reliance of the majority of Member States on fossil fuel imports would reduce significantly, freeing resources for potential investments in development policies and economic modernization [28]. For example, European Union Member State expenditure on fossil fuel imports was €266 billion in 2017 [19].

4. INTEGRATION OF NUCLEAR AND RENEWABLE ENERGY SOURCES

Nuclear–renewable HESs establish an energy network in which nuclear and renewable energy resources are integrated directly in a tightly coupled system or coordinated within a grid balancing area, along with energy storage system(s) to provide resilient energy supply to users in the form of thermal energy and/or electricity [29]. The proposed nuclear–renewable HES designs, configurations and control are optimized to established performance goals, such as maximum efficiency, minimum lifecycle costs, high resiliency and reliability (e.g. to meet an established reliability constraint), and minimum environmental impacts (e.g. emissions, land use, water use). These systems are designed and controlled based on key performance indicators (KPIs), which are defined and evaluated in view of user requirements and constraints. KPIs may include the fraction of energy generated from RESs, lifecycle cost, reliability, safety, availability and environmental impacts.

The configuration of nuclear–renewable HESs will be selected based on available coupling strategies in view of resource availability, demand profiles, and national and regional regulations and policies.

4.1. TECHNOLOGIES: SYSTEMS, SUBSYSTEMS AND COMPONENTS UNDER CONSIDERATION

Nuclear–renewable HESs include multiple generation resources (e.g. nuclear and renewable, although others may also be incorporated) that can be utilized in a coordinated fashion to support various energy demands (e.g. electric, thermal, or various energy products). Integration and operation of these various subsystems should meet target demand load profiles for each of the intended applications. Nuclear–renewable HESs can be configured in various ways, including [29, 30]:

— Single input resource (e.g. nuclear) with multiple output products (e.g. electricity, heat, industrial product);
— Multiple input resources (e.g. nuclear, concentrated solar, wind) with a single product (e.g. electricity); and
— Multiple input resources with multiple output products.

4.1.1. Nuclear technology classifications

Nuclear reactors produce thermal energy that may then be converted into electricity. An NPP is a thermal power station in which the heat source is a nuclear reactor. In a typical NPP heat produced via fission within the nuclear reactor is used to generate steam that drives a steam turbine. The turbine is subsequently connected to a generator that produces electricity. Both thermal energy and electricity may be used to support a variety of applications in industrial, residential and commercial infrastructures. Nuclear reactors are typically classified based on output power, both electric and thermal, as described in Refs [31–41]. Typical categories include large power reactors (>700 MW(e)), medium sized reactors (<700 MW(e)) and SMRs (<300 MW(e)). Within the SMR category there are also very small reactors, also called microreactors. Table 1 provides classification of nuclear reactors by key parameters, such as fuel type, moderator and coolant.

Currently, bioenergy is the largest renewable energy demand in the world, and it is primarily used for space heating and cooking. Hydropower is the largest renewable electricity generator in the world. Wind and solar PV systems market penetrations are growing rapidly and both are poised to provide more energy than bioenergy and hydropower by 2040 [4]. Recent configurations of renewable energy technologies and their integration with electricity grids are described as part of smart energy grid engineering practices [33].

4.2. NUCLEAR–RENEWABLE HYBRID ENERGY SYSTEM COUPLING SCENARIOS

Nuclear–renewable HES coupling is generally classified into four categories:

(a) Loosely coupled;
(b) Multiple products, tightly coupled;
(c) Multiple inputs, tightly coupled;
(d) Multiple inputs/outputs, tightly coupled.

TABLE 1. NUCLEAR REACTOR CLASSIFICATIONS *(refer to the list of acronyms)*

Fuel	Moderator	Coolant	Reactor type
<5% LEU or MOX (oxide)	Light water	Light water	PWR/VVER BWR SCWR
	Heavy water		PHWR
	Graphite		RBMK
	Heavy water		PHWR
Natural U (oxide)	Graphite	Heavy water	GCR (Magnox)
	Graphite	CO_2	HTGR GFR
<20% LEU or MOX	Graphite	Helium	MSR
	None	Salt	GFR
	None	Helium	SFR
	None	Sodium	LFR
		Lead/LBE	

In general, existing energy systems fall into the category of loosely coupled systems in which generators are coordinated via grid operators to meet electricity demand. It is worth mentioning that some cases of tightly coupled systems already exist (e.g. district heating and electricity). Other energy demands, such as those of industrial heat users, are often met by dedicated thermal generation facilities. The introduction of thermal energy networks presents new challenges; although electricity can be transported over long distances to a variety of electricity users, heat can only be transported over short distances without significant heat losses. Hence, tightly coupled systems with multiple users would likely have assets configured around a central location for heat utilization, while electricity output could be transported to users located much further from the NPP. Each of these general classifications is described in greater detail in Subsections 4.2.1–4.2.4.

4.2.1. Loosely coupled

In this coupling configuration, resources could be connected via the grid but can and would be controlled in a more coordinated manner than current grid operations. Control systems would be designed in such way that they would not affect the local control strategies of each energy system. The primary focus is given to the production of electricity, ensuring that grid demand is met at all times. Figure 2 shows all possible energy generation resources that might be included, such as nuclear, wind, hydro, solar PV, biomass (biofuel), concentrated solar power (CSP) and geothermal. A specific selection of these resources would be made in a deployed system, depending on regional characteristics and resource availability. It is anticipated that these systems would also include some form of energy storage (e.g. thermal, chemical, mechanical, electrical) to better accommodate fluctuating net power demands.

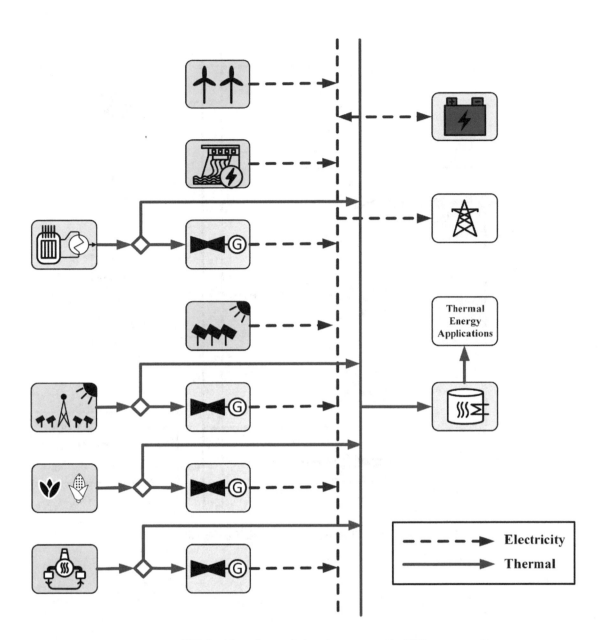

FIG. 2. A loosely coupled nuclear–renewable HES.

4.2.2. Multiple products, tightly coupled

In this configuration, a single generation source is used to support the production of multiple product streams. Thermal energy generated by the nuclear reactor is converted into electrical energy or used directly to support multiple processes. The biomass (biofuel) component, shown in Fig. 3, is produced from feedstock using the heat generated by the nuclear reactor. Thermal energy could be converted into electricity using heat engines, as shown in Fig. 3. Similarly, thermal and electrical energy could be used to produce hydrogen via electrolysis, which would in turn be used in the fuel cells to generate electricity. The hydrogen could also be used in additional processes (e.g. fertilizer production, steel manufacturing) to produce needed commodities rather than electricity; these secondary processes and products are not included in Fig. 3.

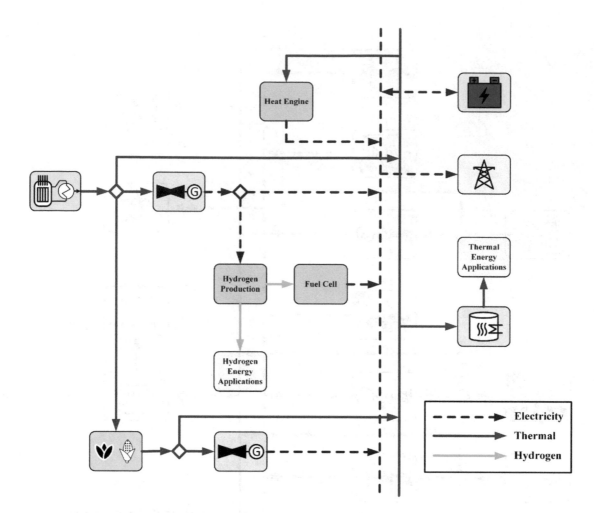

FIG. 3. A multiple product, tightly coupled nuclear–renewable HES.

4.2.3. Multiple inputs, tightly coupled

In multiple input, tightly coupled nuclear–renewable HESs, energy generation resources (e.g. geothermal, nuclear, concentrated solar, PV solar, biofuel, wind or hydropower) are electrically and thermally coupled to generate a single product, electricity. The specific approach to thermal coupling of generation technologies will depend on the subsystem design and will be dependent on working fluids, operating temperature and pressure, and other design specific parameters; such coupling may incorporate a thermal manifold upstream of the turbine and/or thermal storage units. The thermal integration is to consider the impact on the operation and safety of each subsystem. Figure 4 illustrates a general system configuration with all possible resources; produced hydrogen could also be used in the production of additional commodities (e.g. fertilizer, steel manufacturing) rather than for the production of electricity (Fig. 3).

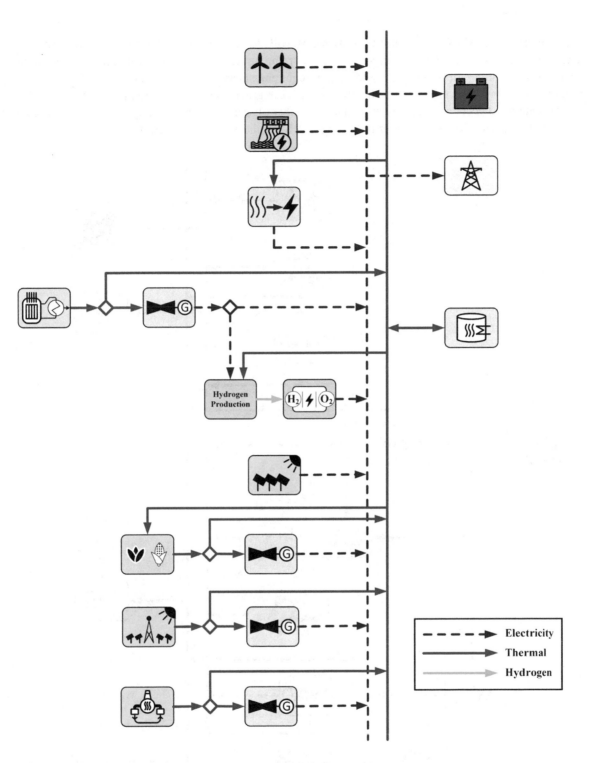

FIG. 4. A multiple input, tightly coupled nuclear–renewable HES with electricity output.

4.2.4. Multiple inputs/outputs, tightly coupled

Figure 5 shows this system configuration where the multiple input resources (e.g. geothermal, nuclear, concentrated solar, PV solar, biofuel, hydro, wind) are connected thermally and electrically to produce multiple energy products. The selected subsystems will depend on regional resources, infrastructure and user requirements. Each implementation of this coupling strategy may have a different configuration and control strategy that will be defined based on user requirements and target KPIs. This type of system requires cocontrol of subsystems in order to be highly responsive to grid demand, have very high energy use efficiency and provide good economic performance under a number of scenarios; produced hydrogen can be used in the production of secondary commodities rather than solely for the production of electricity.

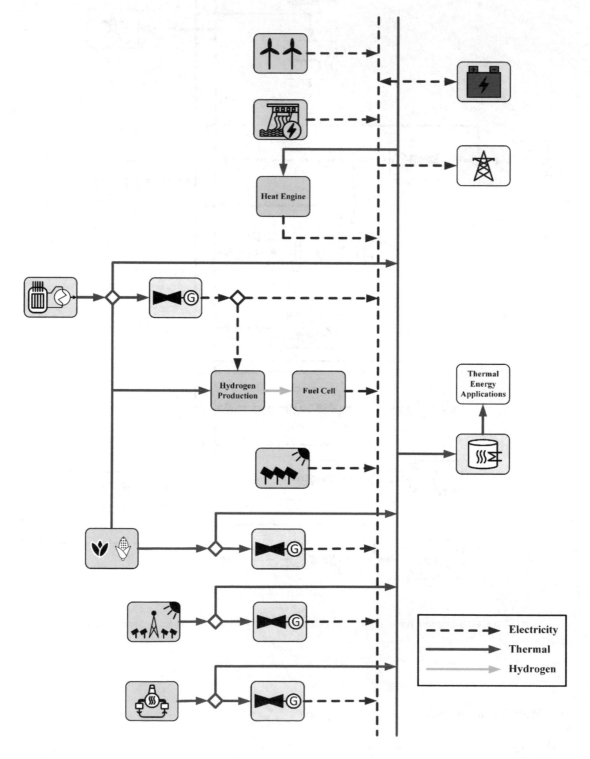

FIG. 5. A tightly coupled nuclear–renewable HES.

TABLE 2. TECHNOLOGY READINESS LEVELS FOR NUCLEAR–RENEWABLE HESs

TRL	Status	Definition	Description
1	Basic research	Observation and reporting	Nuclear–renewable HES concept defined and vetted
2–4	Applied research	Formulation, proof and validation in laboratory environment	Design, modelling, simulation, and small/experimental scale verification and validation of the nuclear–renewable HES concept
5–6	Process and engineering development	Laboratory scale validation and prototype demonstration in relevant environment	Simulation of nuclear reactor heat input with electrically heated and non-nuclear electrically heated components. Integration of renewable components through power converter. Pilot scale demonstration nuclear–renewable HES
7–9	Large scale testing and evaluation	FOAK commercial technology demonstration and commercial deployment	Operation of full nuclear–renewable HES prototype and commercial design and operation of the systems in accordance with the licensing body

4.3. TECHNOLOGY READINESS LEVELS

Technology readiness level (TRL) represents a process where the maturity level of a given technology is assessed quantitatively. It is used to indicate the development of the technology from basic design concept towards commercial deployment on a scale of 1–9, with 9 corresponding to a system that has been deployed commercially. Based on the development progress of a technology, a TRL rating is assigned after evaluation against a set of parameters. This type of assessment is very informative about the degree of technological advancement, the status of the technology, and the time and cost necessary for further maturation of the technology [29]. Table 2 outlines the definitions and concepts for various TRLs for nuclear–renewable HESs [34–37].

The TRL of a specific technology, component, or subsystem does not imply the same readiness level of an integrated system. For example, integration of subsystems that currently operate commercially as independent units, and hence can be designated as TRL 9 when assessed independently, will have a lower system readiness level if they are coupled to operate as a single nuclear–renewable HES. Such system readiness needs to be considered in the development and eventual deployment of candidate nuclear–renewable HESs described in Section 5.

5. NUCLEAR–RENEWABLE HYBRID ENERGY SYSTEM APPLICATIONS

This section provides detailed examples of candidate nuclear–renewable HESs. The purpose of these examples is to enable the reader to identify opportunities that could be considered in national contexts, provide both market and technical information on non-electrical applications and inform the reader about nuclear–renewable HES applications presented here as case studies as follows:

(a)   Heat (Subsection 5.1);
(b)   Hydrogen (Subsection 5.2);
(c)   Water purification (Subsection 5.3);
(d)   Calcination (Subsection 5.4);
(e)   Energy for chemical industry (Subsection 5.5);
(f)   Multiple resources that generate electricity (Subsection 5.6);
(g)   Microgrids (Subsection 5.7).

The first five case studies refer to tightly coupled nuclear–renewable HESs that have a single energy generation source (nuclear reactor) and multiple energy products (electricity and a second product). As such, they interact with the grid and thus the renewable electricity generation is on the grid. Each of these subsections focuses on non-electricity products and how tightly coupled nuclear–renewable HESs could produce those products while meeting the grid's needs for energy and, in some cases, ancillary electrical services.

To the extent possible, each of the five case studies includes descriptions of relevant markets and market trends; a summary of processes available for producing the specified product, including the type of energy needed to support the application; and relevant R&D areas. Finally, each subsection includes analyses that have been reported elsewhere in the literature, where available, to help the reader understand the opportunities and related issues.

The final two case studies (multiple resources that generate electricity and microgrids) focus on other systems that produce electricity exclusively. The first is a hybrid system that increases the efficiency of production of electricity from nuclear energy. Multithermal generation (i.e. electricity generated from thermal energy from multiple sources, such as nuclear and either concentrated solar power or ocean thermal energy) is described. The last case study refers to the potential for nuclear power within microgrids.

5.1. HEAT

The primary product of nuclear energy systems is heat. This heat, or thermal energy, is generally converted to electrical energy to meet grid demand. Similarly, heat is the primary product of other generators, such as concentrating solar power. Nuclear–renewable HESs could provide a unique opportunity to flexibly support both grid electricity demand and heat customers by leveraging assets provided by each generator technology.

5.1.1. Market

Nuclear energy is an attractive option as a complement to renewables to reduce carbon emissions in the electricity sector. In the United States of America (USA), nuclear energy provided 55% of the non-emitting electricity generation in 2018 [38]. Globally, nuclear contributed 10% of the total electricity generation in 2018. Gross heat production in the then 28 countries of the EU was 670 219 GWh in 2017 and 26.5% of this heat was produced by renewables and biofuels (177 599 GWh), while the share of nuclear (1258 GWh) was 0.19% [39]. The combination of renewables with nuclear energy can provide a large fraction of a system's electricity, while minimizing the inefficiencies associated with curtailed generation or energy storage losses. This hybrid configuration can also decarbonize the industrial sector through direct use of thermal energy [40]. Several countries are embarking on studies to determine the technical and economic potential of such configurations. If a carbon price is included in the economic analysis, nuclear–renewable HESs may have lower costs than traditional fossil fired energy systems [41]. Many countries and regions have implemented or are considering implementing a carbon price to curb the emission of GHGs that many believe play a significant role in climate change. This approach provides an incentive to invest in and deploy clean energy technologies that do not emit carbon or other greenhouse gasses into the atmosphere while discouraging the deployment and operation of emitting generation technologies (e.g. coal, gas and oil fired stations) unless carbon capture technologies are employed [42].

Heat from nuclear–renewable HESs is expected to provide numerous advantages, including direct steam utilization via heat exchangers for industry and district heating via heat exchangers to utilize low quality heat [43, 44].

5.1.2. Application

Technical designs for nuclear thermal systems to support district heating and industrial processes have existed since the beginning of nuclear energy development. However, nuclear generated thermal energy does not have a major role in commercial heat markets at present. The prospects for such use are dependent on how and where the heat market demand characteristics can be matched to what nuclear systems can offer. How quickly, and to what extent, nuclear generated thermal energy could capture a portion of the heat market depends mainly on how well the characteristics of nuclear reactors can be matched to the requirements of the heat market in order to compete successfully with alternative energy sources. In this regard, cogeneration and hybrid nuclear plants could provide supply options for district heating and electricity. For medium to large nuclear reactors, electricity is the main product due to the need for reliable baseload power and the limited energy requirements of any single heat market customer. However, the coupling of NPPs with renewable energy systems could provide an opportunity for these plants to support thermal process needs while also directing some fraction, potentially variable, of thermal energy through a steam generator for electricity production to support the grid [45].

Nuclear plant operation can be optimized for electricity market conditions and demand, allowing district heating to be a by-product by utilizing what is traditionally considered waste heat from the power conversion process. The characteristics of the industrial process heat market are different from those for district heating. However, the need for heat transportation and distribution can be similar for both industrial process heat and district heating. On the other hand, industrial process heat users may not be located within highly populated areas, which by definition constitute the district heating market [45].

The interlinkage between energy and food is key when discussing major global trends such as growing population, poverty and climate change. The energy–food nexus mostly concerns the use of energy in the food supply chain, including food production, harvesting, processing, storage, transportation, retail, preparation and cooking. The food sector accounts for some 30% of global primary energy consumption, contributing significantly to fossil fuel consumption and accounting for approximately 20% of total annual GHG emissions, excluding impacts associated with land use changes that result from plant deployment. Renewable energy applications could be possible at different stages of the food supply chain [46]. It could also be possible to use nuclear–renewable HESs for this sector. Deploying hybrid systems for heating is expected to provide benefits that may include alleviating energy security concerns, boosting local productive activities and reducing GHG emissions. The need for heat in the agriculture/food sector is high; for example, much energy is spent on heating greenhouses. Current decentralized heating systems that run on conventional energy sources (gas, woodchips, oil, etc.) could easily be replaced (from a technical point of view) with centralized heating systems. Such a system to distribute heat for agricultural needs is very similar to district heating systems. The difference is that it could be operated at lower temperatures and could also use the low temperature heat derived from the district heating return line [47].

Renewable generators could be combined or coordinated with nuclear systems to efficiently and sustainably produce electricity, or to support district heating, agricultural and industrial process heat applications. As such, nuclear–renewable HESs could be designed to optimally match the characteristics of the thermal heat market.

5.1.3. District heating and heat for agricultural needs

District heating systems have been evolving since the time of the Roman Empire, where heat from remote thermal springs was transported to public baths. Today, those systems exist all over the world and are used for heating of sanitary water and buildings. The heat needs of the latter fluctuate significantly, as they are mostly dependent on external temperatures and thus vary between seasons. Examples of district heating supported by nuclear cogeneration include Switzerland, the Russian Federation, Ukraine, China and Hungary, among others, providing several decades of operational experience [48]. District heating systems with renewable heat sources are a proven technology used worldwide. Figure 6 shows the schematics for a nuclear–solar HES for the district heating and agricultural sectors [56–61].

A nuclear reactor produces the necessary heat for a steam generator. The steam generated is then heated by the working fluid (e.g. molten salt ~560°C [49, 50]) of the solar parabolic collector unit using a heat exchanger. The working fluid (~280°C [49, 50]) is also used to heat the water in the district heating network. The district heating network can also be operated using turbine steam, derived from nuclear reactor and solar parabolic collector units [51]. The district heating system is mainly composed of centralized heat generation, insulated pipes, pumps and heat stations. The medium for heat transport is water, which is heated in the heat exchanger. Heated water is then transported with the help of pumps through insulated pipes to heat exchangers that couple end users. In the heat exchangers, heat is transferred to the domestic water system. Conversely, district heating water is cooled and is then pumped back in insulated pipes to the heat generation heat exchanger, where the process repeats itself. The operating temperature of different district heating systems can vary, as the temperature of supply water ranges from 65–120°C and the return supply ranges from 25–75°C. The sizes (inside diameter) of insulated pipes and pump power are dependent on the difference between the supply and return temperatures (in addition to heat consumption). A small temperature difference implies greater pipe diameters and greater pump power, and vice versa. Higher system temperatures result in greater heat losses. The lower temperature required for district heating is also more favourable for the use of heat from low quality steam derived from the NPP low pressure turbine [51, 52]. Using the return line heat of the district heating network for greenhouses or for food drying and processing purposes can help increase food availability, reduce dependence on fossil fuels, protect against price volatility and diminish harmful emissions from the sector. Furthermore, it can significantly reduce food waste and the efficiency of the process is further increased [47].

5.2. HYDROGEN

Hydrogen production offers a potential opportunity for nuclear–renewable HESs because, as a commodity product today, it has a large market with potential for growth. Today, hydrogen is produced primarily by reforming natural gas using the following chemical reaction, which also leads to the production of CO_2:

$$CH_4 + 2\,H_2O \rightarrow CO_2 + 4\,H_2$$

Other hydrocarbons can be used as the feedstock in place of natural gas, but that option is less common and, in most cases, less economical.

Hydrogen can also be generated by splitting water with electricity and/or heat; the two products that NPPs generate. Replacing natural gas reforming with nuclear driven electrolysis would reduce the carbon emissions associated with hydrogen production, reducing the industrial and transportation sectors' carbon footprint. Hydrogen produced from nuclear heat and/or electricity can also create a new revenue stream for NPPs.

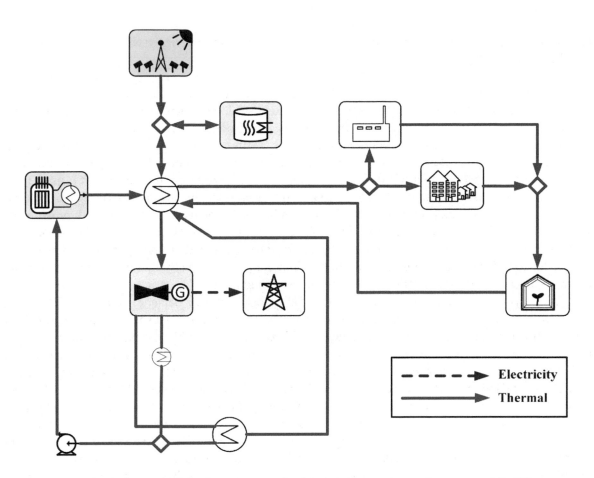

FIG. 6. Nuclear–solar hybrid energy system for district heating and agriculture sectors [47–52].

5.2.1. Applications

Hydrogen is used extensively for a number of applications. Its primary use is in industrial applications, such as oil refining and as an intermediate in ammonia production [53]. Oil refineries currently use hydrogen for cracking, also known as 'sweetening', upgrading heavy fractions to lighter ones, and removing sulphur and other contaminants. Ammonia is produced globally using the Haber–Bosch process that uses nitrogen and hydrogen as feedstocks. It is usually coupled with natural gas reforming, which produces the needed hydrogen, because heat integration improves the overall energy efficiency. Nitrogen feedstock is acquired by an energy intensive air separation plant using compression and refrigeration to produce liquefied nitrogen. Ammonia is then used directly as a fertilizer or reacted to produce an alternative fertilizer such as urea.

Hydrogen is also used to produce methanol, which is a precursor for formaldehyde, acetic acid and other chemicals. Methanol is also used as an energy carrier, either via direct combustion in fuel cells or after conversion to dimethyl ether that is used as an oxygenate to lower emissions from internal combustion engines. Dimethyl ether can also be added to diesel fuel or it can be further dehydrated to produce other oxygenates that are fungible with motor gasoline. In a similar manner, hydrogen can be combined with CO_2 to synthesize formic acid, a chemical that is used to produce laminate materials and may also serve as a hydrogen carrier. More information on organic chemical products is provided in Subsection 5.5.

The hydrogen market is large. For example, as of 2019 global on-purpose hydrogen production is over 60 Mt/year (8.5 EJ/year), produced almost exclusively from reforming natural gas. An additional 40 Mt/year (5.7 EJ/year) is produced as a by-product from other processes (primarily catalytic reforming in petroleum refineries and ethylene production) and is used for processing purposes in the refining and chemical industries. Thus, the by-product hydrogen is usually used internally within the process rather than being marketed [54].

The hydrogen market can grow dramatically because hydrogen can be produced from a diverse range of feedstock (like nuclear and renewable energy used for water splitting) to supply many different end use applications beyond those that are currently established. Examples of potential future hydrogen demands include the use of hydrogen for energy storage, transportation and industrial heating. Because hydrogen is a gas at standard temperature and pressure, it can be stored, although that storage requires geological formations and/or energy intensive compression or liquefaction. If hydrogen can be stored easily it can be an energy buffer that decouples the time energy is produced from the time it is used. It can also reduce emissions associated with the transport and industrial sectors, the two sectors that are difficult to decarbonize. For the industrial sector, hydrogen can be used as both an energy carrier to provide heat when and where it is needed and as a feedstock for chemical reactions. Hydrogen and CO_2 are potential feedstocks for the organic chemical industry (a carbon utilization process), and hydrogen can be used in the production of steel through direct reduction of iron ore [55]. Stationary fuel cell combined heat and power systems can provide electricity and heat/cooling to buildings.

The Hydrogen Council estimates the potential global hydrogen demand in 2050 at 78 EJ/year (550 Mt/year) for electricity storage (primarily seasonal), transportation, industrial energy, feedstock for carbon utilization processes and direct reduction of iron ore, and combined heat and power (CHP) in buildings [56]. The Hydrogen Council identifies transportation as the largest potential hydrogen market at 22 EJ/year (155 Mt/year), a market that has begun to grow in recent years. Between 2014 and August 2019, over 7800 light duty fuel cell electric vehicles (FCEVs) were delivered in the USA [57] and, in 2018, there were 2900 FCEVs in Japan, with hundreds more in Germany, France and the Republic of Korea [58].

Multiple countries have FCEV targets, including the USA (where the state of California has a goal of 1 million FCEVs by 2030 [59]), Japan (200 000 FCEVs by 2025), France (20 000 FCEVs by 2028), China (50 000 FCEVs by 2025) and the Republic of Korea (production target of 81 000 FCEVs by 2023) [58]. Hydrogen utilizing fuel cell options for heavy duty trucks, buses, trains, ferries and aircraft are also being developed [53], with over 21 000 fuel cell forklifts shipped or on order as of 2018 [60]. Fuel cell powered trucks, trains, and ferries are also under development and being featured in early demonstrations worldwide.

5.2.2. Production opportunities

Many options have been proposed or are being developed to produce hydrogen. As mentioned earlier, most of the hydrogen used in the world today is produced from natural gas.

The IAEA [61] identifies six methods in which nuclear energy can be used to produce hydrogen or to supplement other energy sources in producing hydrogen:

(a) Nuclear assisted reforming of natural gas;
(b) Nuclear assisted coal gasification;
(c) Nuclear assisted thermochemical conversion of biomass;
(d) Thermochemical water splitting;
(e) Low temperature electrolysis (LTE);
(f) High temperature steam electrolysis (HTSE).

The first three methods utilize nuclear generated heat (mostly in the form of steam) to reduce the need for fossil or biomass generated heat to perform those conversions. Thus, they are a variation of the chemical industry, as described in Subsection 5.5, and are not described further in this subsection, while the final three are.

Water radiolysis is an alternative hydrogen production technology; however, yields are low and limited research has been conducted on using this technology for industrial hydrogen production [62]. Thus, this option is not discussed further.

5.2.2.1. *Thermochemical water splitting*

Thermochemical water splitting requires water and heat, preferably at a high temperature, and uses a chemical catalyst (or combinations of catalysts) to split the water into hydrogen and oxygen. Ideally, a single step thermal decomposition would be performed; however, that would require temperatures of 2500°C or greater. Thus, multistep processes that require heat at moderate temperatures are being developed [63]. These processes use intermediate chemical compounds to support water splitting and recycle or regenerate those compounds so that they stay within the process.

Thermochemical water splitting is at an early stage of development. Over 350 potential cycles have been identified and most research has been focused on the cycles, as shown in Table 3 [64]. None have been commercialized, although the hybrid sulphur, sulphur iodine, cerium oxide and copper–chlorine cycle has been demonstrated experimentally.

The nuclear–renewable HES interface for all the cycles is heat, which could be transferred as steam or via a high temperature heat transfer fluid (HTF). With a temperature requirement of 550°C, the copper–chlorine cycle has the lowest temperature requirement. The only other two options with temperatures <1000°C are the hybrid sulphur cycle and the sulphur–iodine cycle. Thus, all the cycles require higher temperatures than can be achieved directly from light water reactors (LWRs) without significant temperature boosting. Due to challenges with thermal cycling of materials, these cycles cannot be ramped easily and thus require very stable energy sources [74, 75].

Even without ramping, some of these processes involve strong acids or bases, and materials and pump durability is a key issue; therefore, improving the durability of both the reactant materials and the

TABLE 3. EFFICIENCY AND TEMPERATURE OF HIGH PRIORITY THERMOCHEMICAL WATER SPLITTING OPTIONS (adapted from Ref. [64])

Thermochemical cycle	Potential efficiency (%)	Temperature (°C)
Hybrid copper–chlorine	49	550
Hybrid sulphur	43[a]	750–900
Sulphur–iodine	45[b]	800–900
Cadmium sulphate	55	1200
Barium sulphate	47	1200
Manganese sulphate	42	1200
Cadmium oxide	59	1450
Hybrid cadmium	53	1600
Nickel manganese ferrite	52	1800
Zinc manganese ferrite	52	1800
Zinc oxide	53	2200
Iron oxide	50	2200

[a] Ref. [65] reports potential efficiencies of up to 48%.
[b] Ref. [66] reports potential efficiencies of up to 52%.

equipment that they come into contact with is a key R&D priority, with a focus on the integrity of process components under harsh conditions [67, 68]. In addition, a key R&D area is improving thermal efficiency. Data acquisition and mining are one option being used to address that issue [68]. High temperature reactor design is a second key priority [69]. Additionally, the development of catalysts and separation membranes for high temperature operation in harsh environments is another challenge that requires new research.

5.2.2.2. Low temperature electrolysis

Low temperature electrolysis uses electricity exclusively to split water into hydrogen and oxygen. Alkaline electrolysis has been a commercial technology for decades. It utilizes a pair of electrodes (the cathode and anode) submerged in an aqueous alkaline solution, usually containing either potassium hydroxide or sodium hydroxide. The electrodes are commonly separated by a porous diaphragm that allows hydroxide ions to migrate but prevents the hydrogen that is produced at the cathode from mixing with oxygen that is produced at the anode.

Polymer electrolyte membrane (PEM) electrolysis, also referred to as proton exchange membrane electrolysis, is a newer technology that is related to PEM fuel cells. It uses a polymer membrane that allows protons to cross between the electrodes and prevents mixing of the produced hydrogen and oxygen. This membrane design allows for hydrogen production at up to 20 MPa (thus avoiding or reducing energy intensive compression of the gas). The membrane typically includes a catalyst such as platinum or a platinum group metal to catalyze water splitting, making it a higher cost option than alkaline electrolyzers; however, R&D is targeting cost reductions that will make PEM electrolysis competitive with or have lower cost than alkaline electrolysis [70].

Both electrolyzer technologies can use electricity from any source and have been shown to have up to 69% efficiency based on hydrogen's lower heating value (LHV) [71]. PEM electrolyzers have higher current densities, allowing them to adapt to rapid changes in power that could be caused by a tightly coupled nuclear–renewable HES that is following grid signals. Rapid fluctuations in loads or generation (such as those that might be caused by wind or PV solar generation) can create variability on the grid that requires quick responses from load following entities.

Since the only energy supply to LTE systems is electricity, electricity is the integration point. Within a tightly coupled nuclear–renewable HES, the electricity could be generated from either renewable or nuclear energy, or both. Both alkaline and PEM electrolysis can provide grid services such as flexibility and contingency reserves. PEM electrolyzers can also provide frequency control and voltage regulation because of their ability to ramp very quickly [72]. PEM electrolyzers can further be fully shut off very rapidly and can be put through very rapid cold starts; thus, they do not require a continuous electricity source. In addition, because electrolyzers are modular, the manufacturing process can be scaled. Such scaling is unlikely for thermochemical processes.

LTE systems are a commercial technology, having achieved annual sales of 100 MW (input electricity capacity) in 2017 [73]. At an average electricity requirement of 50 kWh/kg hydrogen produced, that capacity could produce 48 000 kg/day. The larger electrolyzer systems (>1 MW/system) use alkaline technologies, but PEM electrolyzer systems of 1 MW(e) are now also being marketed [70].

Research and development on LTE systems is focused on reducing capital costs with minimal negative impacts on performance. Lower cost materials (e.g. alternative catalysts) and advanced manufacturing methods that can reduce costs are the key focus areas. Improved technologies to achieve moderately high pressure and avoid the need for compression are another R&D area [74].

5.2.2.3. High temperature steam electrolysis

High temperature steam electrolysis is another water splitting technology; it differs from LTE in that water vapour (steam) is split instead of liquid water; thus, lower cost thermal energy can be used to evaporate the water while electricity is then used to split hydrogen from oxygen. The electricity required for splitting water vapour is reduced by the heat of vaporization compared to that required for liquid water. At 750–1000°C, the required electricity input is 35% lower than for liquid water [63], although the electricity saving is reduced at lower temperatures. HTSE can be performed in a solid oxide electrolysis cell (SOEC). The solid oxide electrolyte is a ceramic that conducts oxygen. Proton conducting solid oxide materials are currently being developed and have been shown to be effective for water splitting at reduced temperatures of ~500–700°C. While the kinetics of proton conducting SOECs are slow and thus require a larger cell area to accomplish the same separation efficiencies, the lower temperatures help to avoid several materials durability issues that have an impact on the lifetime of oxygen transporting SOECs and stacks [75].

Since HTSE requires both heat and electricity, both integration points are essential within a nuclear–renewable HES that includes HTSE. Higher temperatures within the HTSE result in higher yields, so high temperature heat is preferred in most situations, but the temperature is limited by the heat tolerance of materials. To avoid thermal stresses, the HTSE stack is optimally kept at a constant temperature. Constant temperature operation requires a reliable heat source, although technologies that allow temperature ramping are under development. Testing of electricity ramping is underway to quantify limitations on ramp rates and frequency.

SOECs have been built and are being tested, and markets are beginning to emerge in Europe and the USA. For example, one system has been deployed in Europe for hydrogen production at a steel mill. Orders have been made from two commercial companies in Europe for commercial technology readiness demonstration [76]. Current R&D focus areas include materials and designs that are durable at high and varying temperatures and improved membranes that increase efficiency [77].

5.2.3. Case studies

Several technoeconomic assessments of nuclear–renewable HESs with electrolysis have been published, including ones with either LTE or HTSE options. In 2005, Ref. [78] proposed cogeneration of hydrogen from nuclear and wind resources using an alkaline electrolyzer. They estimated that the economics of large scale hydrogen production are comparable to those for steam methane reforming with carbon sequestration in Ontario, Canada. Reference [79] considers the nuclear–renewable HES shown in Fig. 7, referring to Pakistan's context. It identifies hydrogen production as a key opportunity for Pakistan to support biofuel refineries and the fertilizer industry by providing a necessary component for biofuel upgrading and the production of ammonia, which is used as a raw material for the production of a variety of fertilizers, synthetic fuels and chemicals. Hydrogen is a key component in ammonia production, which is the backbone of the fertilizer industry — considered to be the largest hydrogen consumer in the world. The use of LTE/HTSE produced hydrogen for several unit processes and unit operations in the chemical industry is briefly discussed in Subsection 5.5.

Reference [80] analyses nuclear–renewable HESs with LTEs that include wind as the renewable energy source and allow for electricity to be purchased from the grid at a higher price than it can be sold to the grid in the USA. They identified opportunities where a nuclear–renewable HES is profitable; however, in the US context, those opportunities require high electricity prices, high natural gas prices, low LTE costs (a purchase price of $100/kW(e) for the LTE system), or drivers beyond economics (e.g. policy). High capacity payments also increase the number of profitable opportunities for nuclear–renewable HESs because they can provide electricity during the hours necessary to receive the capacity payment but produce hydrogen during other hours when its value is greater than the sale of electricity. In cases with very low cost electricity, the study found that the nuclear–renewable HES purchases electricity optimally during periods when the cost is lowest instead of making the capital investment necessary for the nuclear reactor and the steam turbine.

The technoeconomics of nuclear–renewable HESs with HTSE electrolysis have also been analysed. The Japanese Atomic Energy Agency is developing the Gas Turbine High Temperature Reactor 300 (GTHTR300C) for cogeneration; a high temperature gas reactor (HTGR) with the explicit purpose of producing both electricity and heat, making it an optimal reactor choice for nuclear–renewable HESs with HTSE. It is based on a prismatic core reactor and generates power by direct cycle gas turbine with heat as a coproduct [79]. Figure 8 shows this nuclear–renewable HES design with high level controls for

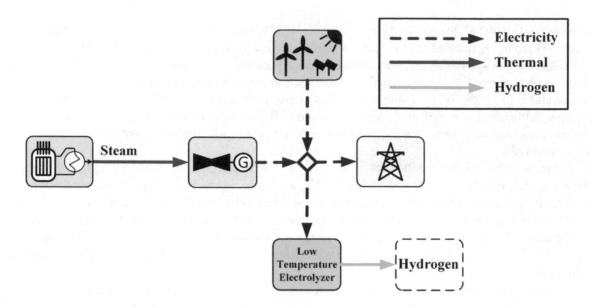

FIG. 7. Block diagram of a nuclear–renewable HES producing hydrogen using low temperature electrolysis.

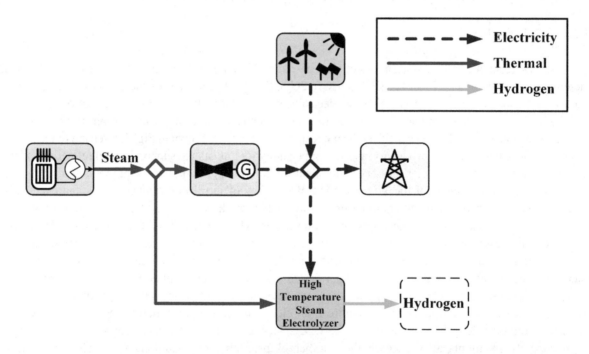

FIG. 8. Block diagram of a nuclear–renewable HES producing hydrogen using high temperature steam electrolysis.

a GTHTR300C, the HTSE nuclear–renewable HES. Preliminary analysis shows that the system has load following capability across both long and short time scales. It also shows that the electricity cost from the GTHTR300 is competitive with that from other sources because the system is simple (with inherent safety and a gas turbine) as well as providing high efficiency electricity production (45−50%) [81].

Reference [80] also analyses the economics of nuclear–renewable HESs with HTSEs. The conclusions are similar to those found for nuclear–renewable HESs with LTEs. The authors reported that although HTSEs have higher efficiencies than LTEs, the increased capital cost did not make the HTSEs an appreciably better fit in nuclear–renewable HESs. They also found that volatile electricity prices impact on the likelihood of HTSE nuclear–renewable HESs being profitable; increased volatility increases the potential for profitability because they can respond to price signals and produce the most profitable product during each hour of the year.

The case for hybridizing existing LWRs to produce hydrogen and electricity is being evaluated for locations where electricity prices are low and volatile due to low cost natural gas and increasing penetrations of wind and solar generation. The analysis of a hybrid process using an HTSE indicates that the optimal configuration involves heat recuperation of the hot hydrogen and oxygen carrying streams to superheat the steam generated using the heat of the LWR, increasing the steam temperature from ~275−775°C. Electrical topping heat is used to boost the steam to 800−850°C before it enters the HTE stacks [82].

Technoeconomic analysis indicates that hybridizing can increase the profitability of the NPP under hydrogen price projections and at a variety of discount rates. Profitability increases primarily due to the ability to maximize hydrogen production during most hours of the year when electricity prices are lower and take advantage of high electricity prices and the capacity market, which has a high compensation for a limited amount of electricity generation. Costs for hydrogen storage and transport from the NPP to markets are highly uncertain and require in-depth analysis to improve estimates [83].

In conclusion, the case studies indicate that nuclear−renewable HESs that produce hydrogen are technically achievable, have the potential to be more profitable than plants that only produce electricity in situations where the electricity price is low and volatile, and there is a robust value for electricity generation capacity. In addition, the ability to store the hydrogen when the nuclear−renewable HES is maximizing electricity production and to transport the hydrogen to markets is a key cost driver. These systems are also more profitable in situations with low electricity to natural gas price ratios.

5.3. WATER PURIFICATION

Access to clean water is fundamental to societal growth, economic development and agricultural productivity. The United Nations Sustainable Development Goal 6 for water focuses on ensuring the availability and sustainable management of water and sanitation for all [84]. As of 2019, the UN estimates that there is still a lack of access to clean water for billions of people globally. More efficient management, distribution and use of existing water resources is critical in addressing water availability and threats to water security, but access to new sources of water through desalination or purification of saline water (brackish or seawater) or wastewater resources may also have significant impact on societal growth. The US Department of Energy has similarly established a Water Security Grand Challenge [85]. This White House initiated challenge establishes a framework to advance transformational technologies and innovation to address the global need for safe, secure and affordable water. Key goals within the Challenge focus on cost effective desalination technologies; transformation of wastewater from the energy sector to a resource; significant reduction in water use by current thermoelectric plants and achievement of net zero water impact for new plants; and development of small, modular energy−water systems for various deployment scenarios (e.g. urban, rural, emergency or disaster response).

Water prices vary dramatically by region, such that no single price point can be established to determine the point at which nuclear driven desalination facilities will be economically viable. The World Bank indicated in December 2017 that "countries need to quadruple spending to US $150 billion a year to deliver universal safe water and sanitation" [86]. The International Desalination Association (IDA) reported that in 2018 Saudi Arabia and Abu Dhabi saw the price for desalinated, potable water drop to US $0.50/m^3 for the first time, with the price reduction primarily attributed to the reduced cost of desalination technologies and increased water reuse. In 2019, the water reuse prices were expected to drop even further to US $0.30−0.40/m^3 [87].

Most desalination facilities currently operate using thermal and electrical energy derived from fossil energy plants. Considering the goals that have been set for reducing carbon emissions and expanding the use of low or zero carbon technologies, it may be opportune to further expand the use of nuclear and renewable energy for desalination applications. Nuclear−renewable HESs could play a significant role in meeting these water needs at a large scale via the use of existing, large scale NPPs or newly built installations, or via modular, transportable systems that incorporate very small nuclear reactor systems. The introduction of water processes via nuclear and renewable energy can simultaneously reduce water use by the facility and support municipal water needs by processing otherwise nonbeneficial water sources.

5.3.1. Prominent desalination technologies

As of mid-2018 the IDA reported in the 31st desalination inventory that more than 20 000 desalination plants were operating around the world with a total installed capacity of more than 97×10^6 m^3 water/day. While the desalination market had been relatively steady since approximately 2015, 2019 was expected to show a surge in development due to increased water demands, decreasing capital and operating costs for desalination, and the need to replace aging desalination facilities with more energy efficient processes [87, 88]. Most of these plants process seawater or brackish water to produce fresh water for potable water use or thermal plant cooling. Prominent desalination technologies fall into two categories: membrane processes and thermal or distillation processes. Membrane processes only require electricity, while distillation requires both heat and electricity. These processes are listed in Table 4

Reverse osmosis (RO) is the most mature membrane desalination process and is employed in approximately 50% of the desalination facilities around the world. RO uses pressure to force water through a series of semipermeable membranes to separate out contaminants and dissolved salts (ions). This process only requires electricity to drive the high pressure pump, which generally operates at 5.5–8.5 MPa. Figure 9 shows a conceptual configuration for the integration of a RO plant within a nuclear–renewable HES.

Two prominent thermally driven distillation technologies are multistage flash (MSF) and multieffect distillation (MED). The integration of a distillation facility is shown conceptually in Fig. 10. In MSF [89], the incoming saline water source is heated to near boiling and pumped through a series of consecutive chambers or stages that have decreasing pressure. At each stage the brackish water or seawater flashes to steam and the vapour is extracted as fresh water; the lower pressure at each stage lowers the boiling point of the water. The salinity of the unflushed portion of water increases at each subsequent stage, such that at each of the series of stages water is removed as fresh water and brine. The MSF plants may have 4 to 40 stages and process on the order of 4000 to 30 000 m^3/day.

The MED is also a distillation process that again requires both thermal and electrical energy input but operates at a somewhat lower temperature than MSF. In this case a steam heat source, such as that which might be acquired from the low pressure turbine in the NPP, is used to process feedwater through a series of evaporators that operate at successively lower pressures. At each stage the saline water source

TABLE 4. RELEVANT WATER DESALINATION OR PURIFICATION TECHNOLOGIES CURRENTLY IN COMMERCIAL OPERATION

Technology	Energy need or coupling option	Temperature (°C)
Brackish water reverse osmosis (BWRO)	Electrical	n.a.[a]
Seawater reverse osmosis (SWRO)	Electrical	n.a.
Multistage flash	Thermal + electrical	90–120
Multieffect distillation	Thermal + electrical	<70
Vapor compression	Thermal + electrical	<65

[a] n.a.: not applicable.

28

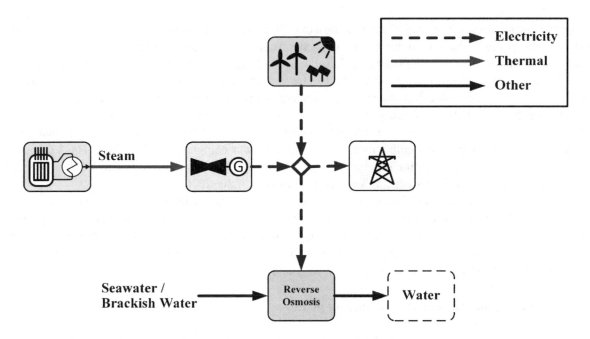

FIG. 9. Conceptual nuclear–renewable HES showing electrical coupling to an RO desalination plant.

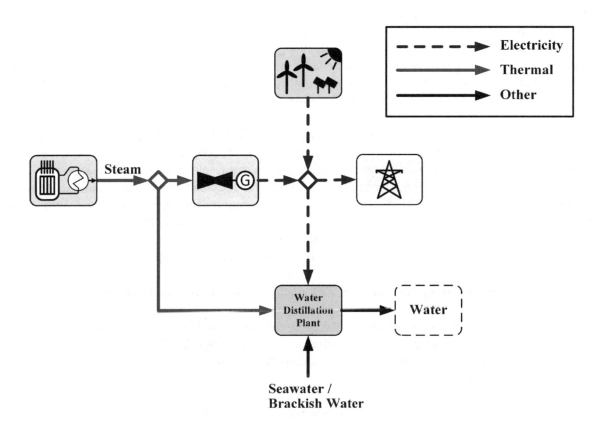

FIG. 10. Conceptual nuclear–renewable HES showing thermal and electrical coupling to a distillation plant (MED or MSF).

is sprayed onto tubes within which steam is flowing. This causes a portion of the feedwater to evaporate off the tube surface; the remainder of the feedwater is collected at the bottom of the vessel, which is then extracted and used as the flow going into the second effect. Steam within the tube condenses and is withdrawn for recovery and use in the NPP. The vapour from the first effect is used to provide the heat source in the second effect. Vapour that condenses inside the tubes from the second effect on is withdrawn as fresh water. Reference [90] provides a detailed description of this process. MED plants may incorporate 8–16 effects and process on the order of 2000–10 000 m^3/day. An additional distillation method that has been used in combination with MED is vapour compression (VC) [91]. The VC process derives heat from the compression of vapour rather than using direct heat from a boiler. VC has generally been applied in small scale systems, processing on the order of 20–2000 m^3/day.

Hybrid desalination facilities draw on multiple desalination technologies to process saline water sources. The Ras Al Khair plant in Saudi Arabia, for example, uses both RO and MSF to process 1 036 000 m^3/day [92]. The site also includes a large power generation component operating at 2400 MW to support the MSF–RO system. Similarly, the United Arab Emirates Fujairah 2 plant processes 591 000 m^3/day using MED and RO, which produce 450 000 m^3/day and 136 500 m^3/day, respectively [92].

5.3.2. Existing nuclear driven desalination plants

Heat and/or electricity from NPPs, coupled with heat/electricity from directly integrated or nearby renewable plants, where applicable, can be utilized for desalination and purification of brackish groundwater, seawater, or wastewater from industrial processes. Integration of these processes provides a pathway for the use of excess energy during times of peak generation from the coupled renewable generators (e.g. during peak solar insolation in the middle of the day) or at times of low electricity demand for a system that also supports grid needs.

Table 5 summarizes currently operating or previously operated nuclear cogeneration plants that produce water either for plant cooling or potable water use. Per the IAEA publication on nuclear cogeneration [23], the global use of nuclear energy for desalination applications has an accumulated experience of 250 years of reactor operation, when tallied across all nuclear desalination units. Except for the desalination facility in Kazakhstan that is now out of service, these units process a small amount of water per day.

As an example of how these systems are utilized in conjunction with an NPP, the Diablo Canyon plant in the USA, operated by Pacific Gas and Electric, processes seawater to support all water use needs on-site, including plant cooling and potable water. The brine that results from the desalination process is rejected and returned to the ocean after being mixed with other rejected water so that local salinity is not measurably increased. The plant is also permitted to provide water for fire suppression to the surrounding area if they are called upon to do so. Pacific Gas and Electric considered increasing the size of the desalination facility to provide potable water to the surrounding communities, but the decision to shut down the plant at the end of the current licences for both reactor units halted the proposed expansion plan.

5.3.3. Case studies for future desalination with nuclear–renewable hybrid energy systems

There are several opportunities under evaluation that would utilize heat and/or electricity from an existing NPP to produce potable water. Two such cases are summarized in this subsection. In many cases developers of advanced reactors and SMRs are also considering coupled desalination facilities, depending on the intended deployment site [94–97].

5.3.3.1.   Case study: nuclear and solar PV to produce electricity and potable water in the US southwest

A 2015 study in the USA evaluated the potential for a nuclear–solar PV nuclear–renewable HES option sited in northeast Arizona. This case included a 600 MW(th) (180 MW(e)) SMR for the nuclear generation and 30 MW(e) solar PV generation. Analyses indicated that such a system could supply 230

TABLE 5. STATUS OF NUCLEAR DESALINATION AROUND THE WORLD [48, 93]

Country	NPP type	Desalination technology	Water capacity (m³/ day)	Status
India	PHWR (Kalpakkam 1,2)	Hybrid MSF–RO	6 300	Operational
Japan	PWR (Ohi 1,2)	MED MSF	(2 × 1 300) (1 × 1 300)	Operational
	PWR (Ohi 3,4)	RO	(2 × 1 300)	Operational
	PWR (Ikata 1,2)	MSF	(2 × 1 000)	Operational
	PWR (Ikata 3,4)	RO	(2 × 1 000)	Operational
	PWR (Genkai 3,4)	MED RO	(1 × 1 000) (1 × 1 000)	Operational
	PWR (Takahama 3,4)	MED–VC	(2 × 1 000)	Operational
	BWR (Kashiwazaki)	MSF	(1 × 1 000)	Not used
Kazakhstan	LMFR (BN-350)	MED, MSF	80 000	Out of service
Pakistan	PHWR (KANUPP)	MED	600	Operational
USA	PWR (Diablo Canyon 1,2)	2 stage RO	2 200	Operational

million cubic metres (61 billion gallons) of fresh water per year, meeting ~88% of the current total water consumption in Phoenix and Tucson, Arizona, using brackish groundwater as the feed source [98]. This initial high level study that considered new generation capacity led to further evaluation among a private public partnership comprising US government laboratory researchers and an NPP owner.

The Palo Verde Generating Station (PVGS) is the largest NPP in the USA. It includes three pressurized water reactors (PWRs), each of which can provide 1.4 GW(e). Arizona Public Service (APS) is the operating owner of PVGS, working alongside six additional plant owners. PVGS serves electricity users in Arizona, New Mexico, California and Texas. This region has seen a dramatic increase in solar PV generation in recent years, which has the potential to impact on the continued steady state operation of PVGS at its full operating capacity. Increasing penetration of solar PV in the region is resulting in excess electricity generation at certain times of day, typically impacting on baseload generation systems. While PVGS currently maintains the ability to operate at its nominal capacity, APS is evaluating the potential future need to operate with flexible power output.

The cooling water for PVGS is currently treated effluent purchased from a regional wastewater treatment plant. APS maintains a long term water resources programme that considers the use of advanced treatment and cooling technologies to reduce plant cooling costs relative to continued use of increasingly expensive effluent. Another option under consideration is to replace some of the costly effluent with brackish groundwater from a regional aquifer, which may require supplemental water treatment to avoid an impact on plant operations. Researchers at the US Department of Energy (DOE) Idaho National Laboratory, working with input from APS, are investigating the option of processing this brackish groundwater, which is unusable directly, using electricity provided by PVGS. This approach would provide secure cooling water supply to PVGS at a stable price, and additional water could be processed for developing municipalities in the west valley of Phoenix. As noted previously, nuclear driven desalination is already employed in the USA by the Diablo Canyon NPP, which operates a seawater reverse osmosis (SWRO) system for both plant cooling and potable water use on the plant site. Although this system is of a smaller scale than what might be envisioned for a RO plant in Arizona, it provides a basis for the plant operation and for the acceptance of a nuclear electricity driven water processing plant in the USA. Figure 11 shows the evaluated nuclear–renewable HES configuration for processing brackish groundwater, in addition to continuing use of some amount of treated municipal wastewater, using electricity produced by PVGS. This nuclear–renewable HES would be categorized as multiple tightly coupled products, as described in Subsection 4.2.2. Although PV is not directly integrated with PVGS, its impact on the PVGS grid balancing area motivates consideration of desalination as a secondary load for PVGS.

Researchers at Idaho National Laboratory developed a methodology to evaluate the potential benefit of increasing the baseload electricity demand by the addition of a desalination plant that would be used to generate cooling water for PVGS from poor quality brackish groundwater. The results of the 2018 preliminary analysis indicate that the proposed configuration may be economically viable if sufficient water is processed to provide for community water needs [99]. A second phase of the study was conducted in direct partnership with the operating utility to identify possible synergies between a larger desalination plant for the production of potable water and various applications in addition to PVGS cooling water [100]. In short, the study indicates that there are feasible combinations of regional RO and PVGS connected RO systems that would support the water purity needs of PVGS and increase municipal water supply, but additional work is necessary to fully evaluate the economic aspects of these systems.

5.3.3.2. Case study: Pakistan nuclear desalination demonstration plant (NDDP)

The Pakistan Atomic Energy Commission successfully commissioned their first nuclear desalination demonstration plant (NDDP) unit at Karachi NPP (KANUPP) in 2010 [101]. KANUPP is a Canada deuterium–uranium (CANDU) type reactor with an installed capacity of 137 MW(e) and has been in commercial operation since 1972. A MED type low temperature horizontal tube evaporator is coupled with KANUPP through an intermediate coupling loop (ICL) [102]. The eight effects included in the MED plant provide 1600 m^3/day of desalinated water with a total dissolved solids content of <5 ppm and a gain output ratio of 6 : 1 [101]. A strategy of low pressure–high pressure–low pressure is adopted to keep the ICL at high pressure compared to the feedwater and primary regenerative heater loop [102]. The purpose of this strategy is to avoid the risk of any radioactive leakage into the NDDP feedwater supply from the low pressure primary loop to the high pressure ICL. Steam extraction from the high pressure turbine at a pressure and temperature of 1.7 bar (170 kPa) and 116°C, respectively, is used in the ICL for nuclear

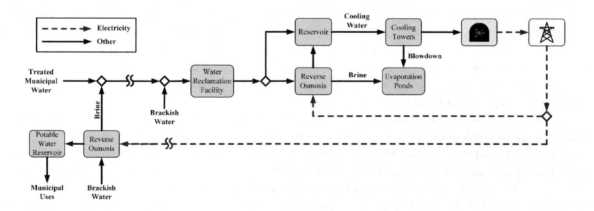

Note: Although solar PV is not directly coupled to the NPP, its presence in the grid balancing area impacts the demand to be met by the NPP.

FIG. 11. Evaluated nuclear–renewable HES configuration for processing brackish groundwater, in addition to continuing use of some amount of treated municipal wastewater, using electricity produced by PVGS.

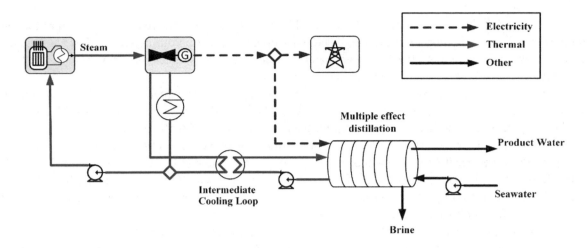

Note: The nuclear desalination facility at KANUPP could be enhanced further by the addition of a renewable, most likely solar, facility.

FIG. 12. Typical process flow diagram for a MED type nuclear desalination facility.

desalination [101]. The ICL consists of an intermediate heat exchanger, recirculation pump and reboiler that produce steam needed for the first effect of the desalination plant. The requirements of feedwater, electricity and process heat are met by the KANUPP facility, while the desalinated water is used for inhouse consumption by the NPP [102]. A schematic diagram of a typical MED type nuclear desalination facility like KANUPP is shown in Fig. 12 [103, 104].

Researchers carried out an assessment study of the KANUPP nuclear desalination project to determine the technical and economic feasibility of various desalination technologies (MED, MSF, RO, hybrid MED–RO and hybrid MSF–RO) for integration with KANUPP [104]. The study shows varying water cost trends for the integrated desalination technologies with an indication of developing hybrid MED–RO or MSF–RO to couple with KANUPP.

Pakistan has vast potential for wind and solar energy [105], and coupling/retrofitting of existing and planned NPPs with renewables (solar and wind) for nonelectric applications such as water desalination could be a good opportunity to address the potable water needs of the country. A nuclear desalination facility at KANUPP could be enhanced further by the addition of a renewable, most likely solar, facility.

5.4. CALCINATION

5.4.1. Global market trends

Concrete is the second most consumed substance on Earth after water; on average, each person consumes three tonnes of concrete per year. Concrete is used as a construction material for buildings and transportation infrastructure and is the final product from cement and aggregate. To produce cement, limestone and other claylike materials are heated in a kiln at approximately 1400 °C; a process called calcination. Today, minerals are predominantly calcined using fossil fired vertical kilns that have replaced earlier, less efficient, horizontal rotary kilns. During the calcination process CO_2 is released both as a result of power generation (burning fossil fuels) and the calcination reaction itself, which is provided for limestone:

$$CaCO3 \ (s) \rightarrow CaO \ (s) + CO2 \ (g)$$

Worrel et al. [106] first estimated that the cement industry, the largest calcination industry, contributes approximately 5% of annual global anthropogenic CO_2 emissions. Ali et al. [107] more recently concluded that in the cement industry roughly half of the emissions result from heat production and half from the calcination reaction itself. The work on CO_2 reduction during calcination has focused on two areas: carbon capture [108–113] and using alternative energy sources that may provide the high temperatures needed for the traditional calcination process.

One alternative energy source considered is CSP. Solar calcination experiments were conducted by Flamant et al. [114, 115], Licht et al. [116], Meier et al. [117–120], and Salman and Kraishi [121]. The listed CSP system concentrates solar radiation for the mineral reaction directly and is thus able to reach or even exceed the high temperatures required for the traditional calcination reaction. Other designs, such as the one used in the recent Low Emissions Intensity Lime and Cement (EU-LEILAC) project [122, 123] or the flameless calcination unit designed by Rheinisch-Westfälische Technische Hochschule Aachen University and Tsinghua University [103, 104], use HTFs such as molten salt or superheated steam that heats minerals for calcination indirectly.

Calcination systems that use HTFs to transfer heat to a mineral feed material reach lower calcination temperatures than systems that concentrate solar radiation directly on a small volume of the mineral feed material. Systems using HTFs can, however, reach much larger throughputs and inherently have the ability to store energy so that energy from the source can vary across time. In addition, various power sources can be used to heat the HTF. Those two aspects make use of an HTF a relevant application for nuclear–renewable HESs.

5.4.2. Production opportunities

Nuclear–renewable HESs use the synergies that are created when coupled systems consisting of nuclear and renewable components are designed to work together [43, 126, 127]. These technical and economic advantages may also be realized in calcination units where the calcine is effectively used to absorb energy; particularly excess energy in the form of heat. This opportunity allows NPPs to continuously produce process heat at their maximum capacity while variable energy production from solar thermal concentrators is captured. Salt storage units allow all of the components in the system to produce at their maximum respective potential. The calcination units are unlikely to ramp their energy use up and down very well due to thermal constraints; thus, energy storage is needed to decouple the energy generation from utilization.

Figure 13 shows simplified schematic overviews of the innovative flameless calcination unit coupled to an HTGR (right) and a concentrated thermal solar power plant (left). The solar power plant was motivated by the Gemasolar power plant recently commissioned in Spain that has a receiver thermal power of 120 MWth [128, 129]. The HTGR was motivated by the HTR-PM demonstration plant currently under construction in Shandong Province, China, that will provide 2×250 MWth [130, 131] for electricity production. One unit of the preconceptual tube in tube flameless calcination system processes some 125 t of mineral feed per day. Depending on the ore that should be processed, two units may need to be employed in a row for complete calcination of the mineral feed. Large calcination plants produce thousands of tonnes of mineral products per day, so it is likely that a number of furnaces will be deployed together. Unlike traditional calcination furnaces that need to be operated 24/7 because of the large amount of calcine in the furnace at one time, the tube in tube calcination system presented here is flexible in its operation, as only relatively small amounts of calcine are in the furnace at one time. This characteristic allows flameless calcination systems to be used as process heat sinks in nuclear–renewable HESs in the same way that other energy industrial applications are, as summarized by Ruth et al. [132]. While standard approaches to calcination have a long operational history, the proposed flameless calcination application for nuclear–renewable HESs remains in an early development phase.

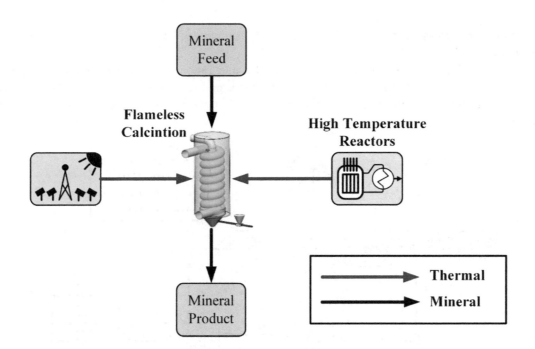

FIG. 13. Schematic overview of the innovative flameless calcination system with CSP (left) and high temperature gas cooled reactors (right).

5.5. CHEMICAL INDUSTRY

Dependence on fossil fuel resources and the risk of negative environmental impacts associated with burning these fuels could be mitigated by adopting low and zero emission energy technologies for electric and nonelectric energy needs. As discussed previously, NPPs can be designed to generate process heat, electricity and feedstock to produce chemical products such as syngas, soda ash, high purity hydrogen, nitrogen, argon, ammonia, methanol and many other chemicals, even at times when RESs such as wind and solar are not adequately available.

5.5.1. Applications

Nuclear reactors provide an excellent source of heat for various industrial applications, including hydrogen, desalination, unconventional oil production and biomass ethanol production. Heat produced by nuclear reactors could be utilized for a variety of applications, such as heating, drying, crystallization (sugar) and distillation in chemical processes, replacing the energy derived from the combustion of fossil fuels. Thus, nuclear–renewable HESs could be used to avoid combustion of fossil fuels and biomass resources, reserving them as feedstock for numerous products in the strategically important chemical industry sector, with the major advantage being reduced emissions.

Nuclear reactors can provide process heat and steam necessary to carry out industrial processes (chemical reactions) and unit operations (physical processes). HTGRs can produce superheated helium, which can then replace the burners in steam methane reforming of natural gas to produce hydrogen, greatly reducing the GHG emissions from this process [133]. HTGRs and molten salt reactors are capable of supplying high steam temperatures. Other designs may also supply high steam temperatures when incorporating heat recuperation of the hot hydrogen and oxygen carrying streams to superheat the generated steam. Additionally, electrical topping heat may be employed to achieve high temperatures.

Figure 14 shows a variety of possible applications of nuclear–renewable HESs in the chemical industry [29].

Table 6 outlines several candidate technologies for integration with nuclear–renewable HESs, with a high level assessment of the TRL or availability of the technology [29, 61, 133–138]. The technologies presented in Table 6 are all in commercial operation, but their implementation in NPPs is at the conceptual stage.

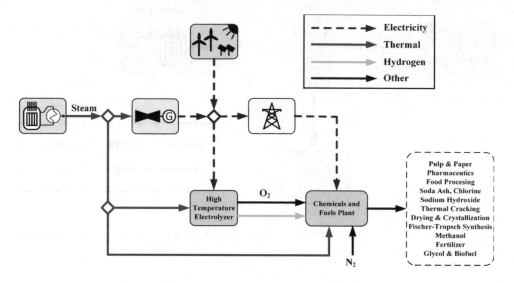

FIG. 14. Examples of applications of nuclear–renewable HESs in chemical industry.

TABLE 6. INTEGRATION AND TRL/AVAILABILITY OF NUCLEAR–RENEWABLE HES BASED CHEMICAL PROCESSES

Technology	Primary coupling	Intermediate coupling	TRL/availability
Syngas based chemicals	Electrical/thermal	Hydrogen	Pilot plant
Pulp, paper and food processing	Thermal/electrical	n.a.[a]	Commercial
Oil refinery and pharmaceuticals	Thermal/electrical	Hydrogen	Conceptual
Biofuel refinery	Thermal/electrical	Hydrogen	Pilot plant
Coal gasification	Thermal/electrical	Hydrogen	Conceptual
Drying and crystallization	Thermal	n.a.	Commercial
Distillation	Thermal	n.a.	Commercial
Ammonia and derivatives	Thermal/electrical	Hydrogen	Conceptual

[a] n.a.: not applicable.

5.5.2. Research and development directions

Some R&D investment is necessary before nuclear–renewable HESs can be deployed for chemical industry applications to ensure rapid, efficient and safe responses to market signals. Chemical manufacturing plants are typically designed to operate at near constant levels; new design schemes that are resilient to time varying electrical and thermal inputs are probably necessary to realize the full potential of nuclear–renewable HESs that support electrical grid demands in addition to chemical processes. Detailed economic models are required to compare conventional and nuclear–renewable HES based chemical industry to evaluate capital costs, operation and maintenance costs, potential product costs and market risks. Study of the environmental impact of the proposed technologies is also essential to characterizing climate change impacts. Furthermore, safety (prevention and mitigation of accidents) and licensing are also to be evaluated for the nuclear–renewable HESs.

5.5.3. Production opportunities

IAEA NES No. NP-T-4.3 [136] discusses the potential application of nuclear energy to chemical industry processing (e.g. petroleum, petrochemicals, hydrogen, steel, iron, aluminium, waste to energy (including plastic recycling)), with a brief introduction to the types of nuclear reactors suitable for various unit processes and unit operations (nuclear process heat reactor design).

The potential nuclear reactor energy delivery source (steam) to support chemical industry applications can be divided into three types: low pressure steam (LPS, <1 MPa), intermediate pressure steam (IPS, 1−10 MPa) and high pressure steam (HPS, >10 MPa) [29]. The requirements for key applications are summarized in Table 7 [29, 139]. In terms of GHG emissions, coal and solid waste gasification provide higher emissions than NPPs. An overview of the HTGR integrated industrial processes for coal and natural gas derived products is provided in the following subsections [135, 138].

5.5.3.1. Coal derived opportunities

The processes involved in conventional coal to liquid, coal to natural gas, and coal to methanol and gasoline plants generate surplus heat to fulfil plant requirements. In a nuclear integrated system, for integration purposes, nuclear generated hydrogen (as described in Subsection 5.2) is required in lieu of direct nuclear heat. By integrating nuclear power and HTSE to support these coal derived processes, GHG emissions could be significantly reduced [135].

Analyses suggest that integration of nuclear electricity and nuclear generated hydrogen would result in increased carbon usage, and reduced coal consumption. As a result, CO_2 emissions would also be decreased for coal to liquid, coal to natural gas, coal to methanol and coal to gasoline applications [135].

5.5.3.2. Natural gas derived opportunities

After detailed review of conventional natural gas to liquid plants, it was observed that there is a significant opportunity for high temperature heat integration. Natural gas combustion in conventional plants can be replaced by nuclear heat provided by an HTGR or other high temperature advanced reactor, as the operating temperature in the primary reformer for conventional plants is 730°C. This small reduction in temperature does not have any negative effect on the performance of the primary reformer [135].

Analyses suggest that replacement of natural gas combustion in the primary reformer with HTGR generated nuclear heat would result in a reduction of natural gas consumption and a decrease in CO_2 emissions in natural gas to liquid and natural gas to methanol and gasoline case studies [135].

TABLE 7. REQUIREMENTS FOR KEY CHEMICAL INDUSTRY APPLICATIONS

Chemical process	Operating temperature (oC)	Other requirements
Pulp, paper and food processing	30–300	IPS
Thermal cracking (oil refineries, pharmaceutical plants, etc.)	300–650	HPS and hydrogen
Inorganic mineral production (soda ash, fertilizers, phosphate, sodium hydroxide, chlorine, etc.)	350–500	HPS, hot gas and molten salt
Biofuel refinery (distillation, torrefaction, pyrolysis, gasification, etc.)	150–1,000	LPS, IPS, HPS, hot gas, hydrogen and molten salt
Organic chemicals (methanol, 1,4 butanediol ethylene/propylene, acetic acid, formaldehyde, resins, hexamethylene diamine, etc.)	150–600	LPS, IPS, HPS, hot gas, hydrogen and molten salt
Coal gasification (syngas, chemical synthesis)	1000–1300	Hydrogen and oxygen

5.5.4. Case studies

IAEA NES No. NP-T-4.2 [61] emphasizes the production of hydrogen and the use of that hydrogen in various industrial processes (e.g. natural gas reforming, coal gasification, thermochemical conversion of biomass). These industrial processes utilize nuclear generated heat to reduce the fossil or biomass generated heat necessary to perform chemical conversions by using HTSE produced hydrogen. Nuclear hydrogen production is described in Subsection 5.2. Coal driven chemical industry and ammonia production are selected as the case studies and are described in the following subsections.

5.5.4.1. Coal driven chemical industry

Global coal based chemical industry is growing significantly in coal rich countries of the world, such as South Africa, the USA, Australia, Indonesia and India. Hence, there is significant potential in the development of low carbon, organic based nuclear–renewable HESs. Hence, it is expected that the penetration of HESs will grow significantly over the next decade [137].

Coal driven chemical industry involves the synthesis of different chemicals to meet societal needs. Carbon obtained from coal gasification serves as a key component in these applications. The water gas shift reaction used for the synthesis of coal derived chemicals results in very high CO_2 emissions. CO_2 emissions can be reduced by integrating nuclear generated hydrogen with carbon obtained from coal gasification to synthesize chemicals with the additional potential benefit of this becoming economically viable [137]. In producing fuel and chemicals in coal based nuclear–renewable HESs, coal serves as a carbon resource while nuclear energy serves as a resource for H_2, O_2, heat and electricity. This type of integration results in enhanced energy and carbon efficiencies and yields an optimum balance between carbon and hydrogen, which leads to reduced CO_2 emissions [138]. Currently, new advanced nuclear plants have 39% integrated thermal efficiency. In comparison, new CCGT plants can reach up to 49%. Advanced NPPs aim at thermal efficiencies significantly above 40%, without considering cogeneration designs. For organic based nuclear–renewable HESs (discussed in Section 5.2.2.3), a coupled HTSE plant could provide H_2, which can then be mixed with syngas obtained from coal/biomass to obtain the required H_2/CO ratio. The O_2 from the HTSE process is then utilized for the coal/biomass gasification process. In this manner, the HTSE plant can serve as a good alternative for conventional water–gas shift reaction units and air separation units [138]. The chemical resources required for coal/biomass gasification, syngas

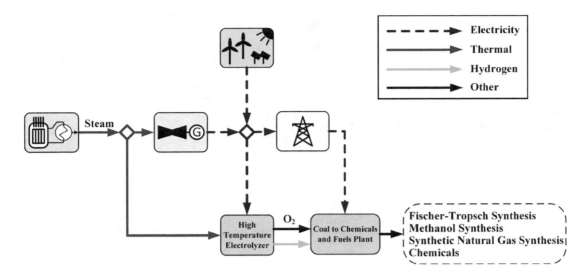

FIG. 15. Simplified energy flow diagram of a nuclear–renewable HES option for coal based chemical industry.

cleaning and sulphur recovery would be the same as those for conventional systems. The nuclear reactor and/or integrated RESs could provide the necessary electrical resources. The coupled nuclear reactor may also provide steam supply to support process and utility heat demands. A simplified flow diagram is shown in Fig. 15 [137].

5.5.4.2. Ammonia production

Conventional production of ammonia is based on the Haber–Bosch process, where the reforming of natural gas with steam produces a gas mixture of hydrogen, nitrogen and CO_2. Nitrogen and hydrogen combine chemically to synthesize ammonia, while CO_2 and other contaminants are removed by the process of absorption before the process of ammonia synthesis:

$$N_2 + 3H_2 \rightarrow 2NH_3$$

Ammonia is a major feedstock required to produce a variety of fertilizers, fuels, chemicals and nitric acid. The two basic raw materials (hydrogen and nitrogen) can be produced at NPPs to replace the dispersed heat market necessary to support the direct raw materials market [139]. The primary advantages of producing ammonia using nuclear energy include low stable operating cost, reduced GHG emissions and the use of readily available raw materials (water and air) rather than expensive fossil fuels. Ammonia production is the largest consumer of hydrogen in the world. Hydrogen produced by HTSE (discussed in Section 5.2.2.3) as a replacement for steam–methane reforming can eliminate the absorption of contaminants in the production of ammonia.

The nitrogen required to produce ammonia can be provided either by cryogen air separation, pressure swing absorption, or burning hydrogen to remove oxygen from air. Thus, coupling of a large scale ammonia plant with HTSE can reduce GHG emissions and provide an economical solution for the production of ammonia and its derivatives [134, 139]. The production and capital cost comparison for nuclear assisted ammonia production found in the literature indicates that, relative to conventional processes, it has the lowest operating cost (10–40% less than conventional) and the highest capital cost (65–430% greater than conventional) [134, 137, 138]. It is predicted that SMRs can reduce this high capital cost by reducing the time and cost of NPP construction and licensing [134]. The integration of renewables (such as solar and wind) with nuclear generation for nuclear assisted ammonia production could further improve plant economics and further reduce GHG emissions. A possible nuclear–renewable HES configuration for nuclear assisted ammonia production plant is presented in Fig. 16 [134].

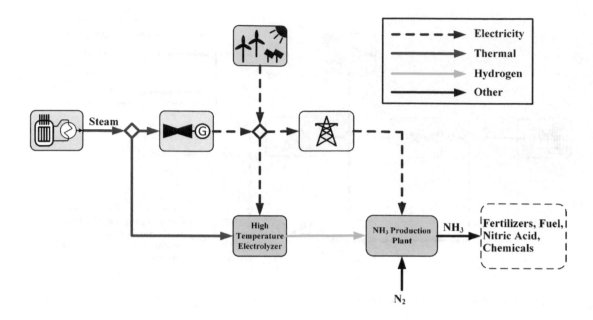

FIG. 16. Simplified energy flow diagram for a nuclear–renewable HES option for ammonia derived chemical industry.

5.6. MULTIPLE RESOURCES FOR ELECTRICITY

Nuclear hybrid systems can be coupled in such a way that they only produce electricity. Electricity presents one of the most practical forms of energy, as it can easily be transformed into other forms. According to the International Energy Agency (IEA) [25], the global demand for electricity is rising and will continue to rise for a period of time. The efficiency of electricity production in nuclear or renewable power plants is being pushed to its limits (especially conventional nuclear). However, process efficiency can be improved for integrated nuclear–renewable HESs, as will be described. In this subsection, two types of hybrid systems that only produce electricity are presented: nuclear combined with CSP and with ocean thermal energy conversion (OTEC). Both technologies are still in the research stage.

5.6.1. Nuclear–renewable hybrid energy system with concentrated solar power

Both nuclear and CSP utilize the Rankine cycle in which heat from steam is transformed into electrical energy in a turbine generator unit. High pressure saturated steam is produced in the steam generator of a conventional NPP. Due to design limitations, this steam has a relatively low temperature and cannot be overheated, resulting in limited plant efficiency and large thermal discharges from the condenser to the environment. In contrast, CSP power tower technology can achieve very high temperatures (250–1000°C) [140], and CSP can generate ultracritical steam temperatures of >600°C [141]. CSP uses mirrors or lenses to concentrate energy from the sun to heat up an HTF, such as molten salt [142]. The heated fluid is connected to a steam generator, which then produces high pressure steam [143].

System efficiency can be increased by combining both technologies. One of the simplest ways to couple the technologies is by installing an additional heat exchanger, such as a solar thermal reheater located between the moisture separator reheater and low pressure turbine (Fig. 17) [144]. In this heat exchanger, heat from CSP further overheats steam from the moisture separator reheater. The proposed system can be installed in existing NPPs or new builds. A second way to couple the two systems is more complex, as it proposes the installation of a superheater before the high pressure turbine with a combination solar thermal reheater located between the moisture separator reheater and low pressure turbine [145].

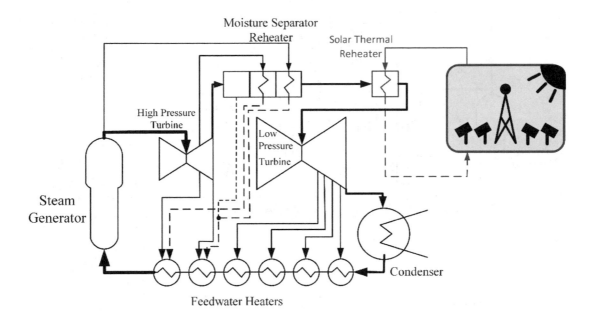

FIG. 17. HES with CSP, reproduced from Ref. [144].

Heat storage tanks installed in the CSP circuit can further improve efficiency and steady operation, recognizing that CSP generation is dependent on both day−night solar irradiation variability and weather.

One advantage of the hybrid configuration illustrated in Fig. 17 is that the power block and associated CSP balancing of the plant is not needed. Elimination of these components can significantly reduce both capital and maintenance costs. A second advantage of this system configuration is increased efficiency. However, there are also some challenges that should be taken into a consideration. During times of low production from the coupled CSP, the benefits of integration may be outweighed by losses due to reduced steam pressure drops in the reheater and superheater. This issue can be mitigated by installing bypasses. Another challenge in defining the system configuration is that the plume from the neighbouring cooling towers may affect the operation of the CSP. Integration of CSP and nuclear generation ensures that there will be sufficient electricity to meet demand at all times. In the event that excess electricity is available from such configuration during times of peak solar production, heat storage tanks already present in the CSP configuration can be utilized to delay electricity generation until solar irradiation is reduced.

5.6.2. Nuclear−renewable hybrid energy system with ocean thermal energy conversion

OTEC operates on the basis of the temperature difference between deep seawater and surface seawater. This process uses an organic Rankine cycle. This cycle is the same as the Rankine cycle in principle, but it uses organic compound as its working fluid rather than water [146]. This enables operation at lower temperatures. The proposed nuclear−OTEC HES [147, 148] uses hot water discharge from the NPP condenser to provide heating in the evaporator (Fig. 18). This water is hotter than the surface seawater; consequently, this hybrid system is more efficient than conventional OTEC. The other benefit of this system is that it extends the suitable regions for OTEC to places with lower seawater surface temperatures and places with shallower deep water. Higher efficiency, shallower deep water intake and combined electrical facilities contribute to the superior economic feasibility of this nuclear−renewable HES compared to conventional OTEC. OTEC is still under development and there are the challenges to overcome in the design of the proposed hybrid systems, such as how to optimize the circulating system and the arrangement of the evaporator at the NPP discharge area [144].

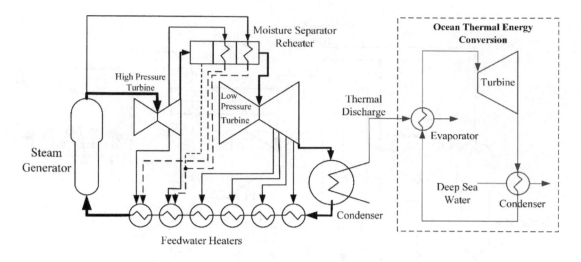

FIG. 18. HES with OTEC [144].

5.7. MICROGRID INTEGRATED WITH NUCLEAR POWER

A microgrid is a small scale electricity generation system at the distribution end with extensive controllability. A microgrid can operate in a grid connected mode (connected with a regional large scale electric grid) or in islanded mode [149, 150]. A microgrid can be fully supported by RESs, but their variable nature suggests a need for integration with a dispatchable source of energy, such as energy storage or flexible nuclear energy, to reliably meet electricity demand. Recent evolution from large scale NPPs to small modular and microreactor concepts provides a viable option for microgrid application [32]. As presented in Section 4, SMRs have an electricity generating capacity of <300 MW(e) [32]. There are a number of challenges to be overcome for microgrids, such as the availability of dispatchable resources at the right scale, lack of long duration energy storage systems in the commercial market, and voltage and frequency regulation. A number of gridscale battery storage systems have been installed globally, and the cost associated with these systems has declined in recent years, but these systems still do not address storage beyond a multihour duration to address seasonal needs. SMRs may be a reasonable option for overcoming these challenges.

5.7.1. Global market and trends

Microgrids are widely used in a number of applications, including, but not limited to, industrial and commercial facilities, transportation and city infrastructure. Microgrids operating in remote communities or facilities are likely to be operated in islanded mode; those connected to a broader regional grid may select the microgrid configuration for enhanced independence or grid resilience. The size of a microgrid may vary, depending on the target load profiles and demand requirements. Small industrial microgrids are commonly rated between 200 kW(e) and 5 MW(e). Utilities can utilize microgrids on feeder lines, with typical sizes from 5–20 MW(e). Power substations might utilize microgrids having capacities that exceed 20 MW(e). In some cases, larger microgrids may be needed to provide larger generation capacity to meet demand profile, such as islands, marine applications and remote cities, where demand can reach 300 MW(e). This case study describes one potential design of an SMR based microgrid and proposes performance indicators, including reliability and grid resiliency.

5.7.2. Regional markets and trends

An SMR based microgrid could support a number of different applications in various regions around the world. These applications may include reliable supply for electrified transportation or marine systems, or supply of heat and electricity to large industrial facilities or remote communities. Regional deployments could include the northern territories in Canada, the State of Alaska in the USA, Greenland (Kingdom of Denmark) and island nations, just to name a few. As many of these regions have expressed a desire to move away from emission generating technologies, such as diesel generators, the market for microgrids based on non-emitting generation technologies is expected to grow.

5.7.3. Proposed microgrid configuration

A microgrid incorporates a number of distributed energy resources (DERs), including energy generators (e.g. renewable, fossil, nuclear), energy storage systems, load management systems and communication networks. As renewable resources are not available at all times, fossil fired (e.g. diesel) generators are often used as backup power to serve the base and emergency loads. However, diesel generators present environmental concerns due to GHG emissions, such that many microgrid users are exploring clean alternatives. An SMR or MMR may be a suitable replacement for fossil fired generators operating within a microgrid [160].

This case study provides an initial assessment of the performance of an SMR in a microgrid relative to a more standard diesel generator. The studied microgrid is based on a tightly coupled HES with multiple input resources and multiple output streams, as illustrated in Fig. 19.

The renewable energy sources in the example case include solar, wind, biomass and hydropower, with dispatchable energy provided by either a nuclear reactor or a diesel generator. Both electricity and thermal demand are considered within the system. Multiple energy storage options are incorporated, including battery banks, thermal energy storage (TES) and hydrogen.

Detailed analysis can be conducted to evaluate and compare microgrid performance when either a diesel generator or an SMR is included in the microgrid; such detailed analysis of the proposed microgrid

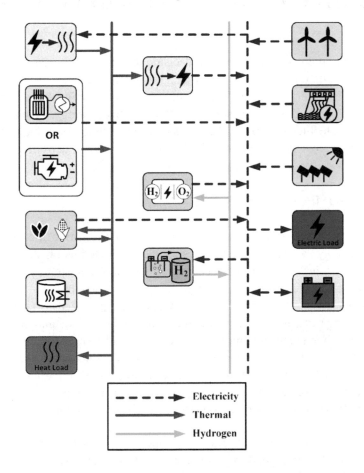

FIG. 19. Microgrid case study schematic.

is beyond the scope of this publication. The analysis approach necessarily entails the definition of key performance indicators and goals and the definition of a control strategy and hierarchy of operations, considering the potential order of operations for both energy generators and users.

The analyses reported in Refs [160, 161] reveal that SMRs could potentially replace diesel generators in the studied microgrid. The SMR based configuration would significantly reduce GHG emissions while reducing energy costs and maintaining reliability of supply. Although SMR based microgrids require additional at-scale demonstration, licensing and siting activities, their potential warrants further development.

Research is still necessary to improve the performance of microgrids. Potential areas for R&D activities include:

— Resilience of microgrids with self-healing and fault tolerant control strategies;
— Intelligent control with prediction capabilities and optimization using artificial intelligence and computational intelligence algorithms;
— Real time monitoring using sensor networks based on improved communication layers; and
— Enhanced SMR designs with autonomous features to support intelligent control systems, with real time control based on microgrid conditions.

6. CONSIDERATIONS FOR NUCLEAR–RENEWABLE HYBRID ENERGY SYSTEM DEPLOYMENT

Many steps should be considered prior to a decision to proceed with the implementation of a new energy system. This section outlines these steps, which range from technical and financial considerations to less quantitative sociopolitical drivers. These steps may entail action by a sponsoring government or government agency seeking wise investment in the country's energy systems, action by research organizations to address technical gaps in the available technologies to achieve the desired goals, or actions by plant designers, owners, or operators in their energy planning activities.

6.1. TECHNO-ECONOMIC ANALYSIS

Technoeconomic analyses are conducted to concurrently determine the technical and economic viability of proposed energy systems. Such analyses can be conducted using steady state modelling tools to first determine the appropriate energy balances within a proposed system, followed by more detailed dynamic analyses to ascertain energy flows in real time in order to optimize the system configuration (e.g. components or subsystems to include), component and subsystem sizing (e.g. capacity of the generation sources, energy storage components and coupled facilities), and then optimization of the dispatch of energy to either electricity production to meet grid demand or to the coupled processes over a simulated operational period. These analyses are conducted within established technical constraints (e.g. maximum/minimum ramp rates or turndown ratios for all of the coupled systems), while also introducing an optimization goal, such as maximum profitability of the nuclear–renewable HES or minimum levelized cost of energy. Examples of such technoeconomic analyses can be found in Refs [83, 151].

As a part of this optimization, the hierarchy of applications is then established. For example, if the nuclear–renewable HES is designed to produce multiple products, what product is considered to be primary? If electricity demand is to be met at all times with an established reliability goal, then electricity production may be selected, regardless of the economic potential of the non-electric applications (e.g. hydrogen or chemical products). Alternatively, the system may be designed to support electricity demand only under certain conditions, whereas the coupled industrial application is to be supported with a specified reliability goal. This hierarchy will impact on the economic evaluation and the potential benefit to the nuclear–renewable HES owner(s); thus, it is considered within the decision framework.

6.1.1. Framework for evaluating the intended nuclear–renewable HES implementation

In order to initiate a technical analysis, one firstly ascertains the intended applications for the energy system, the resources that might be leveraged and the markets to be served. Data on the behaviour of the renewable resources (i.e. wind or solar potential, even if these have not yet been built in a region), electricity demand as a function of time (over a full one year period) and projected demand growth, and energy markets are critical in evaluating the feasibility of a nuclear–renewable HES. Further considerations and data requirements are elucidated in the following subsections.

Many steps should be considered prior to a decision to proceed with the implementation of a new energy system. This subsection outlines these steps, which range from technical and financial considerations to less quantitative sociopolitical drivers. These steps may entail action by a sponsoring government or government agency seeking wise investment in the country's energy systems, action by research organizations to address technical gaps in the available technologies to achieve the desired goals, or actions by plant designers, owners, or operators in their energy planning activities.

6.1.1.1. Regional energy demand and opportunities

Key considerations in selecting potential energy generation and use options, and necessary datasets to support evaluation of those options, are summarized as follows:

(a) Resource potential for renewable generation sources (e.g. wind, solar, hydro):
  (i) What renewable resources are currently installed in the region, or have a reasonable expectation to be installed?
  (ii) Are there geographical features present to support energy system options (e.g. is the regional geography supportive of using pumped hydro for energy storage that may be integrated within the nuclear–renewable HES)?
  (iii) Are the natural features appropriate for the selected technologies (e.g. is elevation available to support pumped hydro? Is sufficient wind available?)?
  (iv) If built, what is a reasonable expectation for availability or capacity factor for each resource?
  (v) Are data available at sufficient fidelity for the renewable energy source as a function of time (e.g. are the available data at a five minute or hourly resolution? Are a full year of data available to ensure understanding of seasonal behaviours?)?
(b) Energy and electricity demand in the intended deployment region:
  (i) What is the current and estimated future baseload, peak and time dependent electricity demand?
     — At what time resolution are these data available?
     — What is the anticipated growth in electricity demand over the intended lifetime of the energy system under consideration due either to population or industrial growth (e.g. what will the demand be in 40 or 60 years)?
  (ii) What thermal energy demands are present in the region currently, and what is the potential for growth of industrial markets if a high quality, reliable energy source is available?
(c) Regional resources present:
  (i) Is water available in the region to support cooling of thermal plants?
  (ii) In regions of water scarcity, is there a brackish water, seawater, or wastewater resource available?

— Note that such a resource could be used to support the production of water for plant cooling and/or for potable water use in municipal regions.

— If additional, cost effective water resources were made available what is the anticipated impact on population and industrial growth, which could further increase energy demand?

(iii) Are there available sources of coal, natural gas, or other feedstocks to support the production of commercial commodities?

6.1.1.2. Suite of technologies to be considered

After the regional parameters described in Subsection 6.1.1.1 have been characterized, the full suite of technologies that could be utilized in meeting the various demands should be identified. The capacity of the components or subsystems may be determined by conducting specific analyses focused on determining the best utilization of each of these options to meet the established technical, economic, or environmental performance goals.

(a) Nuclear energy system:
   (i) Selection of a candidate nuclear energy technology firstly considers the power level/range necessary to support the intended applications. Evaluations might consider options such as those described in Section 4: MMR, on the order of one to tens of MW(e); SMR, <300 MW(e); or large scale reactor, on the order of one GW(e).
   (ii) Selection of a candidate reactor technology should then consider the temperature required to support the coupled application(s):
      — Water cooled technologies produce steam at a temperature on the order of 300oC. This output steam temperature might be increased via chemical heat pumps or electrical heating, although the impact of such a configuration on the overall plant efficiency is considered in the analysis. Selection of water cooled technology with temperature boosting may reduce risk, as these reactor technologies are already available in the commercial market.
      — Advanced reactor technologies under development may offer mid-range (i.e. 500–700°C) or high temperature (i.e. 900–1,000°C) output, but there may be additional project risk if a novel technology is selected.
(b) Renewable energy source:
   (i) Selection of a renewable energy source considers the resource potential, as identified in Subsection 6.1.1.1, and any constraints that are applied, such as maximum facility size due to land availability.
   (ii) Optimization analysis using an approach such as that described by Epiney et al. [152] can then be employed to select the appropriate size renewable facility to be integrated in the nuclear–renewable HES. The estimated time dependent availability of the renewable resource is applied as a part of the dynamic system analysis and optimization.
(c) Coupled energy use facilities, if included in the nuclear–renewable HES:
   (i) If the nuclear–renewable HES includes a coupled energy use facility, such as an industrial plant or chemical production facility, the appropriate size of that plant is determined. Such selection of plant capacity can be made as a part of the multivariable optimization problem.
   (ii) The size or capacity of the coupled energy use facility considers the current and potential future market for the produced commodity, feedstock availability and necessary transport infrastructure, and infrastructure necessary to transport the produced commodity to the customer.
(d) Energy storage:
   (i) It may be necessary to include some form of energy storage in the plant design to ensure operational stability as energy is dynamically allocated among the various coupled energy

users or to ensure that the energy demands are met with an established level of reliability (e.g. stored energy may be used to support peak electricity demand).

    (ii)   Evaluation should consider multiple viable energy storage options as part of the system optimization, including thermal energy storage, chemical energy storage (i.e. hydrogen) and electrical energy storage. The dynamic analysis toolset applied in the optimization should employ subsystem models with the appropriate performance constraints (e.g. charge and discharge rate, storage capacity and duration) to ensure adequate selection of storage systems. Note that the optimization toolset may include multiple storage technologies in the analysis, but the results could select zero capacity for one option while building out a second option.

(e)   Financial parameters:

    (i)   System design and optimization analysis includes an estimate of capital costs for each plant, as well as other fixed and variable costs (e.g. operations and maintenance, fuel).

    (ii)   Analysis should also estimate the lifetime of each subsystem that will be coupled. Each subsystem is likely to have a different operational lifetime; therefore, it is necessary to assess when a specific component may need to be replaced over the lifetime of the nuclear–renewable HES to ensure that the capital costs are properly accrued over the full nuclear–renewable HES lifetime. For example, a nuclear plant has a typical lifetime of 60 years (40 years for the initial licence plus 20 years for licence renewal); however, a coupled industrial plant might only be expected to operate for 30 years. Hence, the evaluation includes the cost for building a second coupled plant 30 years into the project to ensure use of the overall capital investment to the greatest extent possible.

    (iii)   System designers incorporate the desired system performance goals associated with selected economic parameters (e.g. profitability index, levelized cost of electricity (LCOE), internal rate of return) in parallel with assessment of environmental impact measures and adherence to the reliability requirements for each of the coupled systems.

Once the analyst has selected options to include in the optimization and has applied the appropriate financial data for each of these options, the selected dynamic analysis toolset may then be applied to determine optimal component and subsystem sizes based on a selected goal function (e.g. maximum profit, minimum LCOE). This analysis should determine whether or how to include externalities (e.g. environmental goals, assurance of grid reliability) in the optimization, particularly if those externalities are captured via economic incentives or disincentives (e.g. renewable or clean energy credits or a carbon tax).

6.1.2. Technology readiness to implement

Each technology selected for a nuclear–renewable HES may have a different technology readiness that will impact on the overall deployment timeline and risk associated with project completion. Standard definition of the TRL was introduced in Subsection 4.3. Even if all subsystems are commercially available when the project is being considered, if the specific nuclear–renewable HES configuration desired is not currently deployed then the readiness of the integrated configuration may not be fully commercial. Considerations pertinent to the readiness of each technology are summarized as follows:

(a)   Subsystems:

    (i)   Are any of the selected subsystems currently deployed commercially, either at the intended location (i.e. an existing plant or facility will be repurposed for hybrid operation) or at another location?

    (ii)   If not, what is the relative TRL and timeline associated with achieving commercial availability? What investment of both time and funding is necessary to raise the TRL to support deployment? If selected, what is the associated build time for the facility?

— A large scale LWR is commercially available, but the time to build such a facility is significant due to the magnitude of the facility and the need for significant site preparation, limited supply chain for large scale components, and on-site build and assembly. An advanced reactor, on the other hand, may be in the conceptual design stage or undergoing design certification. This lower TRL concept may take a number of years to achieve commercial availability, but it may also have a shorter anticipated build time because of new manufacturing and assembly approaches, reduced physical size and/or reduced site preparation requirements. Hence, it may not be immediately apparent which technology selection might result in the shortest time to completion. These trades impact on the type and extent of project risk that would be associated with the technology selection and should be taken into consideration before deciding on a path forward.

(b) Interconnections, sensors and control systems:

   (i) Are there commercial off the shelf solutions for interconnections (e.g. heat exchangers, valves), sensors and control systems, or will new technologies be needed?

      — If possible, early stage implementations may consider the use of high TRL technologies to avoid long development time horizons while R&D, technology development and demonstration, and/or regulatory review are conducted for more innovative solutions. Advanced sensor technologies or semiautonomous control systems may fall into the latter category.

      — Example: behind the grid electrical integration of a current fleet LWR with a hydrogen plant (see Subsection 5.2) or RO desalination (see Subsection 5.3) may not achieve the highest efficiency possible for such systems, but these demonstrations can provide a foundation for later thermally integrated technologies or advanced reactor implementations by improving public perception, providing a much needed commodity and demonstrating safe operation of operating nuclear, renewable and chemical plants in close vicinity, if these plants are co-located and connected behind the grid interconnect.

When evaluating new technologies to be incorporated in the nuclear–renewable HES that are not yet commercially deployed, the considerations summarized below should be evaluated systematically, with project risk taken into consideration at each step:

— If there is not a commercially available solution to meet a specific need, what new technologies are required, and what is the associated TRL of those technologies?

— Is the work necessary to enhance the TRL currently being conducted by national laboratories, academia, or industry?

— What data are missing to fully deploy the technology (e.g. performance, reliability, lifetime)? Does the technology exist at a laboratory or bench scale but require scale up for implementation?

— What further steps are necessary to move this technology forward?

— Can technologies be demonstrated in a non-nuclear environment, or is a nuclear environment required?

6.2. REGULATORY REVIEW

Implementing any new energy project requires engagement with multiple regulators. Such engagement should be initiated early in the project to increase awareness of pending requests for review and to identify any potential roadblocks early in the design and feasibility stages. For a nuclear–renewable HES, such engagement will necessarily include the relevant nuclear regulator for the intended deployment location, environmental review boards, industrial review boards, the electricity regulator and the electricity market operator. Key considerations and steps for each of these regulatory bodies are summarized below. Each step in the review process can add significant time to the schedule for project implementation. The steps and timeline for each of these reviews may differ, depending on the intended deployment location and the relevant regulatory bodies; hence, estimated timelines for each step are not included in the summary.

(a) Nuclear regulator:
  (i) The typical timeline for review by the nuclear regulator may differ for each nuclear–renewable HES under consideration, as some options may include modification of existing facilities or may incorporate newly designed systems. Different review types may include:
      — License amendment for existing plant (electrical integration);
      — License amendment for existing plant (thermal integration);
      — New reactor plant;
      — Design certification;
      — Construction licence (assuming use of a certified design);
      — Operating licence (assuming use of a certified design and facility has been constructed); and
      — Combined construction and operating licence (optional regulatory review path in some countries).
  (ii) Significant data and analyses are supplied to the nuclear regulator for each review and approval stage. Such requirements may be quite different for the proposed hybrid operation versus production of electricity. Engaging the regulator early in the nuclear–renewable HES design and evaluation process will support in outlining the remaining R&D efforts that will precede design certification or licence amendment submission.
(b) Environmental impact assessment:
  (i)   Determine the types of analyses and characterization studies required for the nuclear plant, and the associated timeline for completing these studies;
  (ii) Determine the types of analyses and characterization studies needed for the renewable plant, and the associated timeline for completing these studies;
  (iii) Determine the types of analyses and characterization studies needed for the coupled energy use facility (e.g. chemical plant), and the associated timeline for completion;
  (iv) Determine whether there is any overlap in the environmental review requirements (e.g. characterization of the seismicity of the intended deployment location) and ability to conduct the necessary assessments in parallel;
  (v)   Engage with the appropriate environmental review groups to determine whether the planned nuclear–renewable HES will trigger the need for environmental reviews that differ from those for standalone systems.
(c) Electricity regulator:
  (i) In some electricity markets, electricity generators are not permitted to vary when they supply electricity to the grid based on economic considerations from the plant owner perspective. For example, a plant owner/operator may be permitted to reduce power sent to the grid to avoid selling electricity at a loss (negative electricity prices), but they may not be permitted to divert energy to the production of alternative products simply to increase revenue from the sale of that product versus electricity. Such operation could be considered market manipulation. These guidelines will likely vary by location and country and should be evaluated for each case.
  (ii) Coupling of energy users electrically (versus thermally) may be fairly straightforward, but it may not be readily apparent how such coupling will be treated by the electricity regulator from an economic perspective. If coupled behind the grid interconnect for the energy system, the coupled energy user may be considered house load. Alternatively, some regulators may require that standard retail rates be applied regardless of whether the coupling is via the grid or behind the grid interconnect. The latter scenario would suggest that locating the coupled plant anywhere on the grid (within the same balancing area) would be equivalent to colocation within a HES.

(d) Electricity market operator:

    (i) Determine the electricity market structure for the intended deployment locations under review. Operating the system in regulated versus deregulated markets could have a significant impact on the economic viability of the proposed system.

    (ii) Determine the bidding times for electricity generators operating within the intended market (e.g. bids in a day ahead differ from real time markets and each impacts on flexibility requirements).

    (iii) Determine whether the nuclear–renewable HES will operate in a capacity market in which it may be compensated for the ability to ensure adequate power supplies for the grid.

6.3. STAKEHOLDER ENGAGEMENT

The stakeholders include all those who are involved; key organizations (government, operator, regulator), as well as any individual or group with an interest or who feels that they may be affected by the project. Engaging stakeholders early in the selection, design and development of energy systems, including both nuclear and renewable technologies, can be critical to the success of the project [153]. From a technical perspective, such engagement would include designers, owners and operators of the types of subsystems in the intended nuclear–renewable HES to ensure that systems are technically compatible and to ensure that the stakeholders for each subsystem are comfortable with the proposed integration. Host community stakeholders are expected to be engaged before decisions are taken or significant investment is made to allow the community to be part of the process. This approach is being applied in many countries for the siting of nuclear related facilities and is often referred to as 'consent based siting' [154]. Key considerations that may arise in identifying appropriate stakeholders for engagement are summarized below. Engagement needs may vary in each country or region due to cultural or legal differences.

If there is an existing nuclear facility, the public sentiment associated with that facility should be evaluated and understood. If no facilities are currently deployed in the region, one should ascertain if anything with regard to public sentiment specifically prevented previous deployment of nuclear energy facilities. Based on community outreach activities and the overall public sentiment, decision makers should assess the potential for the technology (i.e. the selected nuclear–renewable HES and its subsystems) to be accepted by the host community. If the host community is not expected to be receptive, a public outreach and information campaign may be required prior to moving forward with the nuclear–renewable HES. It is necessary to determine which groups of stakeholders in the impacted host community are statutory in the decision making process. Bringing those decision makers or decision influencers into active discussion early in the process of designing the nuclear–renewable HES will help reduce the overall risk associated with host community acceptance of the project. Deploying a new energy system, such as a nuclear–renewable HES, may have a significant beneficial impact on jobs in the region. Identifying this benefit and sharing it with the affected host community could have a substantial impact on willingness to accept a new type of energy system. Finally, the incorporation of non-emitting energy technologies could have a significant impact on public health. In fact, the province of Ontario, Canada, where 90% of electricity is provided by non-emitting sources (58% nuclear, 22% hydro, 8% wind, 2% solar in 2017), estimates annual savings in health costs of CAD $4.4 billion relative to scenarios in which fossil based technologies were to provide the equivalent amount of electricity [155].

6.4. POLICY AND GOVERNMENTAL CONSIDERATIONS

Policies established by local or national governments can have a significant impact on the selection and deployment of energy systems. As of May 2019, 194 States and the European Union, representing almost 97% of global GHG emissions at that time, signed the Paris Agreement, agreeing to reduce environmental emissions in order to limit the increase in global temperatures. While some countries lack an overall national energy policy, these goals are often reflected at the state, province, or city level. Many utilities are also moving to similar goals for clean energy production and use [156]. The Intergovernmental Panel on Climate Change projects a 2.7°C increase in global temperatures by 2040 if global GHG emissions continue at the 2017 rate. As a result, many new goals and standards have been established to deploy clean energy technologies. While some of these declarations focus solely on renewable generation technologies, others specifically include nuclear energy in recognition of nuclear energy's role in the reliable generation of both thermal energy and electricity without emission of CO_2 or other greenhouse gases. These policies can be implemented in a number of ways.

Some countries or regional governing bodies have chosen to provide compensation to non-emitting generators or to apply penalties to emitting technologies. These incentives may come in the form of feed-in tariffs to incentivize deployment of clean energy generators, or they could be applied as a tax on carbon emissions that is applied to emitting technologies (generators or industrial applications). Other incentives may come in the form of preferential selection of clean energy products by end users. For example, a company using hydrogen in its process may preferentially purchase hydrogen made via non-emitting processes versus the more standard steam methane reforming, even if it comes at a higher cost. Declarations have also been made by large electricity users, such as data centres, to purchase non-emitting electricity. With large companies such as Google making such declarations, it is anticipated that more large scale energy users will follow suit in the coming years.

Other policies that could impact on capacity expansion choices may include regional energy (or commodity) trade policies (e.g. across state/province/national borders) or policies that limit industrial operations to use of a domestic supply chain or limit the export of the resultant product. Macroeconomic factors such as economic output, unemployment, inflation, savings and investments should also be considered, as they impact on economic performance and are closely monitored by governments, businesses and consumers.

6.5. OWNER PERSPECTIVES

As discussed in Subsection 6.1, plant owners, on both the generation and use sides, will have a unique perspective on how a selected nuclear–renewable HES might be designed and deployed. In some countries, the energy plants are owned by governmental entities, whereas plants are privately owned in other countries. Nuclear–renewable HESs present a unique challenge in that owners of coupled energy use facilities may have requirements and considerations that are inconsistent with the coupled generation systems. For example, many industrial processes require constant or near constant heat to operate efficiently and economically. If the nuclear–renewable HES is responding to variable grid demand, the heat sent to the coupled thermal energy user could be at risk of dropping below the desired level. Hence, additional generation technologies may be needed to ensure constant supply of heat, such as a synthetic fuel fired unit. Although such an addition would introduce additional capital investment and environmental emissions, it might be justified to ensure that the coupled process is maintained at the desired operating level. Mismatch in construction times for coupled supply and use facilities could also introduce unwanted risk for the individual plant owners. This ownership model can have a significant impact on the owner's willingness to take on the technical risk of operating the system in a hybrid, multiproduct mode versus production of electricity alone or the economic risk associated with investing in a large scale project and the expected time before a return on that investment is achieved. Owners should also consider the intended plant lifetime, which for any nuclear system would be expected to be on the order of 60 years or more. The market for industrial products could change dramatically within that time period and, as noted previously, the expected lifetime for a coupled industrial facility may be significantly different from the nuclear plant operating period. Hence, the ability to incorporate a new energy user into the nuclear–renewable HES at some point may be of significant interest, such that flexible interconnection and licensing may be desirable. Finally, the business model for owners of a nuclear–renewable HES may be significantly different than for a generator that produces electricity alone. In some cases a nuclear plant may have multiple owners, with one serving as the 'operating' owner, or there may be multiple owners associated with the individual subsystems for the overall energy system because of the significant differences in the expertise necessary to operate and manage a nuclear, renewable, or industrial facility.

Insurance associated with the overall coupled facility is to be considered by the plant owner(s). Insurance became the downfall of one planned cogeneration system in the USA in 2005. In this case, a process steam user, Cargill, Inc., approached the Omaha Public Power District about purchasing process steam to support the processing of food products. The US Nuclear Regulatory Commission found no licensing, security, or safety issues, but the nuclear insurance company was unable to resolve issues associated with a very small probability of tritium migration from the nuclear plant to the ultimate food products. If an additional intermediary loop or heat storage had been incorporated into the proposed system design, it could have mitigated this risk. Such actual or perceived risks are eliminated from nuclear–renewable HESs by design to ensure that plant owners do not fail in their projects because they are unable to secure insurance.

6.6. VENDOR PERSPECTIVES

Many vendors will be involved in the development and implementation of a nuclear–renewable HES. Nuclear vendors have traditionally only considered the application of their technologies to the production of electricity. Application of the reactor plant for alternative applications may be perceived to have an impact on warranties associated with plant components, such as the reactor fuel, control rod drives, or steam generators, to name a few. New applications of nuclear technology may also impact on the supply chain, such that new designs appropriate to the HES application are developed to nuclear quality standards. If the market for such designs is not perceived to be significant, vendors may not be willing to make an effort to develop these components. Vendors should be engaged in the system design phase to avoid such challenges in deploying the planned hybrid system.

7. GAPS

At the time of writing, there was no tightly coupled nuclear–renewable HES in the world. In order to achieve a fully operational tightly coupled nuclear–renewable HES, researchers and scientists, as well as technology providers, need to work together with industry and governments to study and resolve obstacles to ensure the smooth deployment of nuclear–renewable HESs. This section describes both the technical and non-technical gaps and challenges, and possible ways to address them with practical planning to ensure the smooth and successful implementation of coupled nuclear–renewable HESs at different scales.

7.1. OVERARCHING NEEDS

7.1.1. Human capital

The required level of safety for nuclear–renewable HESs should be at least comparable to that for current standalone NPPs. For loosely coupled systems, the safety and operation of the NPP will be minimally affected by the hybridization unless the renewable electricity share is significant. Operator training would be the same as that for a standard NPP. Meanwhile, to guarantee safe and reliable design, operation and maintenance of a tightly coupled nuclear–renewable HES, adequate education and training of operators is extremely important. If the renewable electricity share is significant or beyond some level in a grid, the NPP operator should also take into account the variability of the coupled dynamic renewable generation source.

The required operator training will be dependent on the manner of hybridization. For thermal storage based hybridization, the associated technologies will be quite conventional extension of well known technologies. However, a hydrogen based nuclear–renewable HES should be combined with new and additional subsystems and the system operation and control will be much different than for a standalone NPP. Therefore, it is extremely important to establish nuclear–renewable HES specific training and education course for operators and maintenance staff. For operator training, a new expanded simulator system would be necessary. Relevant training courses should also be developed for the coupled system.

In the training of operators and maintenance staff for the nuclear–renewable HES, the effects of interconnected systems are expected to be considered. A transient or malfunction in a subsystem may propagate to the other system through the connecting subsystems. It will be important to limit the intersystem impacts to support system safety.

7.1.2. Markets and grid regulation

An important motivation for nuclear–renewable HESs is to enable renewable systems to overcome their inherent variability and provide reliable and dispatchable electricity through system integration. At the same time, the integrated system can support improved flexibility of the NPP to load variations in order to maintain resilience while maximizing plant revenue. Both nuclear and renewable energies can be utilized optimally through nuclear–renewable HESs. Storage of surplus energy via electrical, thermal or chemical means, or redirection of surplus energy to coupled industrial processes, further improves technical and economic system efficiency. Such configurations will affect the interaction of nuclear–renewable HESs with the electricity market, and grid regulation may be managed very differently than current markets. Therefore, it is important to ensure that nuclear–renewable HESs are capable of both active load following and frequency control operations to be competitive in future energy markets, and designing such flexibility into the system from the start will offer better performance than later modifications made to system design.

Due to the variable nature and increasing deployment of renewable generators, nuclear–renewable HESs will be required to actively load follow and provide frequency control for the grid system. In conventional NPPs, load following operations have traditionally been limited, and most of these NPPs are not designed for large window power manoeuvring, such as daily 100−50−100 operations [180]. In addition, the required power ramp-up rate due to the expanded renewables is actually significantly higher than current NPP capability. On top of that, active frequency control in conventional NPPs is not easy. Therefore, it is very important to make sure that hybrid nuclear systems are able to do both active load following and frequency control operations for competitiveness in the future electricity market.

If the hybrid system is to produce chemical products in addition to electricity, the system should be able to respond to the associated market environments, and the operation of the NPPs should be flexible, depending on the market needs and conditions for the associated chemical products.

System maintenance will be also be important for nuclear–renewable HESs to maximize system utilization. Conventional NPPs undergo a periodic shutdown for refuelling and maintenance, approximately every 12–24 months depending on the specific reactor design. Consequently, the coupled system may be called on to run solely on the renewable generator while the nuclear subsystem is under maintenance. In this regard, it may be necessary to consider multiple small modular nuclear systems that can be utilized optimally to minimize the planned shutdown time in a harmonious way with the coupled renewable system. It may also be desirable to synchronize planned maintenance of the coupled systems (e.g. hydrogen production facility) with the nuclear refuelling outage.

7.1.3. System integration

Existing nuclear and renewable energy systems adhere to significantly different design standards and requirements, leading to challenges in deploying integrated nuclear–renewable HESs. The same may be true for coupled industrial processes. In particular, nuclear systems are required to adhere to more stringent safety requirements than renewables. However, in order to integrate two different subsystems optimally, it is necessary to standardize the design principles and safety requirements such that the hybrid system meets the necessary safety and performance requirements. The interface design will be crucial for the safe and economical operation of nuclear–renewable HESs. Such interfaces should also allow coupled subsystems (e.g. renewable generator, industrial process) to operate under conventional regulations while nuclear standards are applied to the reactor and associated interfaces. This approach minimizes overconservative designs that only serve to increase cost but maintains rigorous safety requirements.

In a tightly coupled nuclear–renewable HES, system integration can be accomplished through a common medium or component (e.g. heat exchanger or thermal manifold) shared by the coupled systems. In some cases, a tertiary coupling loop may be necessary to ensure isolation of radiation under postulated accident scenarios. The safety of the coupled system will be strongly dependent on the coupling scheme, and it is important to conduct an overall safety evaluation that considers all potential accident scenarios. Such scenarios would include those that are standard for electricity generating NPPs (e.g. loss of coolant accident) and any additional scenarios that could be initiated by coupled systems (e.g. failure or shutdown of a coupled industrial process, leading to sudden loss of load).

If the coupling method employs standard heat exchanger designs or common thermal energy storage media (e.g. solar salts), the associated technologies may be fairly mature (high TRL). However, it is important to note that the integration of technologies or subsystems with high TRLs as independent systems does not guarantee a high TRL for the coupled system. Analysis of the coupled system should consider material compatibility, operational compatibility, pressure and temperature limitations, ramping limitations and potential for radiation to cross boundaries between the systems (e.g. in the event of a steam generator tube rupture). Nuclear–renewable HESs that are only coupled via electrical interconnection avoid many of these concerns, although colocation of subsystems may still impact on the overall nuclear–renewable HES operation due to the introduction of new accident scenarios (e.g. the impact of an industrial plant explosion on nearby nuclear and renewable subsystems). Thermally coupled systems may require the introduction of a tertiary loop that isolates the nuclear system, preventing any potential contamination of coupled subsystems under off-nominal conditions in the nuclear system. Although such configuration could reduce operating efficiency, the reduction in risk far outweighs any efficiency losses.

To ensure safe and reliable operation, the nuclear–renewable HES control system should be autonomous or semiautonomous. The complexity of nuclear–renewable HESs, which make real time decisions to produce electricity or products from the coupled industrial process based on significant amounts of data on renewable generation, electricity demand and predicted future variations in these data sets, mean that it is crucial to employ a 'smart' control system with fewer humans in the loop to ensure safe, competitive commercial operation.

7.1.4. Safety implications

7.1.4.1. Mutual safety impacts

The safety of integrated subsystems can be affected by the interface design. Any potential for one subsystem to impact on another coupled system presents additional safety considerations in a hybrid system. When coupling a steam based nuclear system with other generators or non-electric energy users, for example, measures may be necessary to ensure that the steam used to drive coupled processes is free of any radioactive materials, and that no radiation can be released to the environment. For example, it is possible that radioactive tritium or other fission products could migrate from the nuclear reactor to the steam based secondary system (this could occur under a scenario in which there is a steam generator tube rupture in parallel with a fuel failure). It is prudent to eliminate the potential for migration of radioactive materials via specific design choices, possibly through the introduction of a tertiary loop, as discussed previously.

It is also to be noted that operational thermal margins are likely to be different in the two coupled subsystems. In particular, it is likely that the thermal margin will be smaller in the nuclear subsystem than in the renewable subsystem. In the middle of a transient, none of the subsystems are to hamper the functions and operational thermal margin of the other subsystems.

7.1.4.2. Site preparation

In the case of thermally coupled systems, the coupled systems should be physically located as close to each other as possible to minimize energy losses and maximize system efficiency. However, in the case of hydrogen or chemical production plants, the nuclear system should be clearly separated from the hydrogen or chemical system to ensure that any postulated accidents in the coupled industrial plant (e.g. hydrogen explosion) cannot result in an adverse impact on the safety of the NPP. This consideration establishes a design constraint that may hamper system efficiency; for example, greater separation between two thermally connected subsystems implies larger heat loss.

In general, nuclear systems are located far from large cities and customers due to safety concerns, and nuclear sites should undergo rigorous evaluation to assess potential seismic conditions, ground water impacts and potential impact on native flora and fauna, among many other considerations. Meanwhile, it is common for large renewable systems to be sited close to end use consumers and cities. In order to simplify and reduce the cost of delivery of nuclear–renewable HES products, the coupled system should be sited as close as possible to the end use. Therefore, SMRs or future advanced designs may be preferred for nuclear–renewable HESs because their small physical size, potential to have smaller exclusion zones and passive safety systems allow them to be sited much closer to consumers.

7.1.4.3. Bounding accident scenarios

In a thermally coupled nuclear–renewable HES, both nuclear and renewable subsystems may provide thermal energy to a common energy storage system. Stored thermal energy may be converted to electricity or other product forms. In this case, malfunction of the common thermal storage may be a new source of accident scenarios.

If the nuclear–renewable HES is coupled through H_2 production or other chemical products, postulated new accidents would include hydrogen explosion or possible chemical reactions in the hybrid system. If the hybrid system produces chemicals such as synfuels or biofuels, the release of such chemicals may also present new possible accident scenarios.

Electrically coupled nuclear–renewable HESs are loosely coupled systems. In this case, new accident scenarios are unlikely. Nevertheless, new accident situations could arise if the coupled subsystem produces chemical products. In this case, chemical reactions by the products in the colocated subsystem are to be considered as possible accident initiators.

7.2. SPECIFIC TECHNICAL NEEDS

7.2.1. Retrofit of existing nuclear power plants

NPPs have traditionally been used as baseload power in many countries, with the exception of countries such as France and Germany, which have extensive experience in flexible operation of NPPs due to large fractions of their grid electricity being supported by nuclear or other baseload generators. More recently, the electricity market has been forcing rigid baseload NPPs to follow load or conduct part load operations as the variable renewable electricity share increases in countries such as the USA. It is expected that future energy systems will need load following capability to be competitive in the electricity market. Existing NPPs are loosely coupled with renewable resources via the electricity grid, and they are requested to compete with other resources while contributing to the grid stability. Reference [157] describes a dynamic net electricity demand that results from high renewable penetration, with a specific focus on solar PV generation. In the State of California's grid in the USA, this impact is often referred to as the duck curve, in which net demand and electricity prices are significantly depressed at midday due to high solar generation, followed by a rapid increase in demand in the evening hours as the sun goes down and grid demand increases at the end of the working day. Introduced by an increasing fraction of grid electricity being met by variable renewable energy when it is available, this dynamic net demand behaviour adds stress to other generators on the grid that are to accommodate these changes.

For existing NPPs to be more competitive and more harmonious with renewable resources, many plants are to operate with flexible power output, while others are considering the production of non-electric products to increase flexibility without reducing revenues. In some cases, the NPP balance of plant may require modifications to support load follow operations. Advanced reactivity control technologies may also be necessary, including constant average coolant temperature [158–161]. Where allowed by the governing regulatory body, NPPs can also be upgraded to support active frequency control, optimizing the reactor core and fuel designs.

In conventional NPP load following operation, the maximum power ramp rate is ~5%/min, which is lower than the required ramp rate to accommodate high renewable shares. For safe and reliable load following and frequency control operation, the operational thermal margin of the current fuel designs needs to be enhanced by adopting advanced cladding and/or fuels such as accident tolerant fuel technologies to maximize the power ramp rate [162]. More proactive NPP power manoeuvring will enhance nuclear–renewable synergies in the future. Recent studies show that ideal load following operation can be achieved if the soluble boron can be completely removed in PWRs [163, 164]. In so called passively autonomous power manoeuvring, both load following and frequency control operation can be accomplished without active manoeuvring of the control rods. New burnable absorbers could also be developed to eliminate the need for soluble boron in PWRs.

Another way for existing NPPs to respond to the increasing penetration of variable renewable electricity is to utilize surplus electrical or thermal energy to produce other valuable products, as described in this publication. In these alternative approaches, the required system retrofit or refurbishment is dependent on how the additional products are produced [83, 151, 161].

As discussed in Section 5, non-electric energy users may be coupled electrically, thermally (via heat exchanger), or via offtake of main steam from the NPP secondary side. The latter option would require a steam extraction system to be added to the main steam line, designed to respond automatically to the variability of the renewable electricity. In this kind of system retrofit or refurbishment, it is critically important to make sure that the safety and performance of the plant are not affected in any way. Table 8 summarizes the potential retrofit options.

TABLE 8. RETROFIT TECHNOLOGIES FOR EXISTING NPPs

Technology	Rationale
Improved balance of plant	Better load following operation
Advanced reactivity control	Better load following operation
Accident tolerant fuel	Higher thermal margin and flexibility
Soluble boron free or reduced boron PWR	Higher safety and flexibility
Hydrogen production and desalination	Cogeneration

7.2.2. Greenfield implementations

As discussed previously, SMRs or other future advanced reactors may be appropriate for future tightly coupled nuclear–renewable HESs, particularly for microgrid applications. If SMRs or other future advanced reactors are coupled with renewables for the hybrid system, siting near large cities may be possible due to their exceptionally high level of passive safety and regulatory approval for reduced exclusion zones. Many SMR designs are under development in several countries, and it is anticipated that some of them will be deployed in the near future. The advent of SMRs or other future advanced reactors offers new opportunities for the flexible operation of a nuclear plant relative to a single large reactor. A multimodule facility offers the opportunity to dedicate some modules to electricity alone and others to support coupled processes, with some perhaps indicated as 'swing' modules to respond to variable demand as needed. Additionally, a 50% reduction in power output could be accomplished by wholly turning off half of the modules, rather than reducing a single large module to half of its rated capacity. This module based operation also overcomes the challenges associated with power reductions near the end of the reactor fuel cycle, as experienced by large scale reactors; in a multimodular plant, each unit is at a different stage within its fuel cycle and, hence, different modules can be called on to respond to the need for flexible output as a function of their operating history.

SMRs or other future advanced reactors can be equipped with thermal storage to support more flexible and economical operation [165]. Thermal storage can also be utilized for coupling with renewable resources, such as concentrated solar power. In this case, it is important to design a high performance thermal storage system in view of the intended use of the stored energy and necessary storage duration.

Building a nuclear–renewable HES requires the development of design and analysis methods in order to ensure that the integrated system meets required safety and performance requirements. The relevant licensing authority is also expected to establish extended evaluation processes and capability for a nuclear–renewable HES. To minimize the costs of the hybrid system, standardization and modularization of the components and subsystems will be highly valued. In particular, the construction period for the nuclear–renewable HES should be shortened via modularization and factory fabrication and the system cost can be reduced substantially.

Reliable control technologies are to be designed and demonstrated for overall control of tightly coupled systems, taking into account all state variables of the subsystems and components. In addition, the variability of the renewable resource can be assessed using advanced modelling and simulation techniques to facilitate system utilization and reliability.

In a tightly coupled nuclear–renewable HES, the nuclear system may operate over a long period of time, and planned outages for refuelling and maintenance should be minimized. In this regard, a long life fast reactor concept has a unique advantage due to its long refuelling cycle of 10 or more years, although the plant would still require routine shutdown for maintenance, potentially for a shorter time period than that required for refuelling. Fast reactors are also advantageous in that thermal efficiency can be higher as a result of the higher operating temperature of the coolant [166]. Advanced large scale LWRs can also be coupled effectively with renewable resources in a closer, synergistic way. Innovative technologies currently under development for LWRs, such as advanced technology fuels (also referred to as accident tolerant fuels), may be introduced to further improve safety and the operational margins for these new applications. Table 9 summarizes the contents of this subsection.

If NPPs perform routine daily load following operations, the amount of energy production during a fixed period will decrease, and the capacity factor will be reduced, relative to current baseload operation. Recent studies show that the electricity production cost tends to increase while revenue decreases for NPPs performing load following operations [160, 161, 167]. Meanwhile, the electricity price is highly time dependent and the electricity price of the more resilient NPPs can be made higher by actively contributing to the load variations and stabilization of the electricity grid [168]. Consequently, it is expected that the competitiveness of NPPs may be maintained by optimally combining cogeneration, load following operations and energy storage.

TABLE 9. TECHNOLOGY REQUIREMENTS FOR COMPETITIVE NUCLEAR–RENEWABLE HESs

Technology	Rationale
SMR and future advanced designs	Near consumers due to passive safety, reduced exclusion zone
Heat storage	Simple; improves efficiency
Standardization and modularization	Reduced cost and construction time
Big data and artificial intelligence	Optimal utilization of renewable resources
Small fast reactors	Long life operation and enhanced safety
Heat pipe cooling	Simple, compact core design
SCO_2 power conversion	Compact size demonstrating high efficiency
Advanced large LWRs	Enhanced safety, operational resilience, operational flexibility

TABLE 10. TRL OF TIGHTLY COUPLED NUCLEAR–RENEWABLE HESs

Energy sources	Coupling method	Possible outputs	TRL
Nuclear–wind	Electrical	Electricity, hydrogen, heat	Proof of concept or prototype
Nuclear–solar	Thermal or electrical	Electricity, heat	Proof of concept or prototype
Nuclear–solar–wind	Electrical	Electricity, heat	Proof of concept
Nuclear–geothermal	Thermal	Electricity, heat	Proof of concept
Nuclear–hydro	Electrical	Electricity	Demonstration [169]
Nuclear–biomass	Thermal	Electricity, biofuels	Proof of concept
Nuclear–wind–natural gas	Electric or thermal	Electricity, chemicals, synfuel	Proof of concept
Nuclear–other	Electrical or thermal	Electricity, hydrogen, heat	Proof of concept

7.2.3. Coupled system technology readiness level

The technology readiness level of a nuclear–renewable HES is strongly dependent on the TRL of each subsystem and the coupling and operational schemes, and it can be affected by many other attributes of the hybrid system, including the maturity of the design methods. In the case of loose coupling through electricity, the TRL is largely dependent on each subsystem, and the TRL should be relatively high.

Given that operating tightly coupled nuclear–renewable HESs do not exist, it is difficult to determine the TRL of this technology in a systematic way. Nevertheless, the suggested TRLs for various nuclear–renewable HESs are described in Section 4. Table 10 shows the three TRL categories: 1–3, proof of concept; 4–6, prototype; and 7–9, demonstration in operation. The nuclear–hydro hybrid system is considered to be mature, since pumped hydro energy storage (PHES) technology is already in operation in many countries [169]. Regarding the nuclear–wind and nuclear–solar concepts, it is well known that NPPs are actively involved in stabilization of the dynamic electricity grid via load following and frequency control operations in France, Germany and other countries via loose coupling of technologies within a grid balancing area.

7.3. NON-TECHNICAL GAPS AND CHALLENGES

In addition to the already mentioned various gaps, there is an essential non-technical gap for nuclear–renewable HES development. Various kinds of nuclear–renewable HESs are being proposed by research organizations, reactor developers and utilities that own and operate NPPs. Additionally, there is growing interest among renewable energy experts in some countries, such as the USA, in a viable means to introduce a larger fraction of clean generation technologies across all energy use sectors (i.e. electricity, industry, transportation). In other countries, the nuclear–renewable HES idea does not resonate with renewable experts and a large hurdle will need to be overcome to demonstrate nuclear–renewable HESs. This difference in perception regarding nuclear–renewable HESs should be minimized to support synergistic development of these concepts. In order to persuade renewable experts and others to accept and support nuclear–renewable HESs, it is critical to show that they are mutually beneficial to both nuclear and renewable generators in the long run. Both the research community and technology developers should increase their efforts to engage with and inform the general public about the potential for coordinated use of nuclear and renewable resources to provide clean energy to all energy use sectors. Support from the public for nuclear–renewable HES concepts will be a cornerstone for clean energy, facilitating their implementation.

8. CONCLUSION

Two principal options for low carbon energy include renewable and nuclear technologies. To date, these generation options have been considered primarily as independent contributors to the electricity grid. Potential synergies between these energy options, and the advantages of their coordinated use to support energy needs across the electricity, industry and transportation sectors, have not been fully explored to date. Nuclear–renewable HESs, whether they are tightly coupled via thermal interconnections among generators and energy users, or loosely coupled solely via electricity, are designed to leverage the benefits of each of these sources to reliably meet energy demands at an affordable cost to consumers. This publication explores many potential energy use options that could further amplify the benefits of both nuclear and renewable energy, including desalination, hydrogen production, district heating or cooling, coal to liquids conversion and process heat applications in the chemical industry, among others. While the true value of these multi-input, multioutput energy systems remains to be demonstrated, research to date suggests that nuclear–renewable HESs may play a key role in meeting future energy demands in a flexible and resilient manner, while supporting established sustainable development goals.

REFERENCES

[1]   INTERNATIONAL ENERGY AGENCY, World Energy Outlook 2014, IEA, Paris (2014).

[2]   UNITED NATIONS, "Transforming our world: the 2030 agenda for sustainable development", A New Era in Global Health, Springer, New York, NY (2018).

[3]   INTERGOVERNMENTAL PANEL ON CLIMATE CHANGE, Global Warming of 1.5 °C, IPCC, Geneva (2018).

[4]   INTERNATIONAL ENERGY AGENCY, World Energy Outlook 2019, IEA, Paris (2019).

[5]   GOSPODARCZYK, M.M., FISHER, M.N., IAEA 2019 Data on Nuclear Power Plants Operating Experience, IAEA, Vienna (2019)
        https://www.iaea.org/newscenter/news/iaea-releases-2019-data-on-nuclear-power-plants-operating-experience

[6]   WORLD NUCLEAR ASSOCIATION, World Nuclear Performance Report 2019, WNA, London (2019).

[7]   NATIONAL RENEWABLE ENERGY LABORATORY, Annual Technology Baseline: Electricity, NREL, Washington, D.C.,
        https://atb-archive.nrel.gov/electricity/2019/index.html

[8]   IEA and OECD/NEA, Projected Costs of Generating Electricity, 2020 Edition, Paris (2020).

[9]   NUCLEAR ENERGY AGENCY, ORGANIZATION FOR ECONOMIC COOPERATION AND DEVELOPMENT, Technical and Economic Aspects of Load Following with Nuclear Power Plants, NEA/OECD, Paris (2011).

[10]   DOLLEY, S., Exelon Generation cuts output at four Illinois nuclear units, two back at 100% power, S&P Global Platts, London (2018).

[11]   MCMAHON, J., Nuclear operators scramble to make reactors flexible enough for new energy economy, Forbes (Apr. 2018).

[12]   ELECTRIC POWER RESEARCH INSTITUTE, US DEPARTMENT OF ENERGY, Program on Technology Innovation: Fossil Fleet Transition with Fuel Changes and Large Scale Variable Renewable Integration, EPRI/DOE, Palo Alto, CA (2015).

[13]   EUR ORGANISATION, European Utility Requirements for LWR Nuclear Power Plants, EUR, Lyon (2016).

[14]   ORGANIZATION OF THE PETROLEUM EXPORTING COUNTRIES, Annual Statistical Bulletin, OPEC,
        https://asb.opec.org/index.php/interactivecharts/gas-data

[15]   ENVIRONMENTAL INVESTIGATION AGENCY, Henry Hub Natural Gas Spot Price, EIA, London (2019).

[16]   KWON, H.M., MOON, S.W., KIM, T.S., A study on 65 % potential efficiency of the gas turbine combined cycle, J. Mech. Sci. Technol. **33** (2019) 4535.

[17]   CONGRESSIONAL RESEARCH SERVICE, Financial Challenges of Operating Nuclear Power Plants in the United States, CRS, Washington, D.C. (2016).

[18]   UNITED NATIONS, The Paris Agreement (2015),
        https://unfccc.int/sites/default/files/english_paris_agreement.pdf

[19]   EUROPEAN COMMISSION, A clean planet for all — a European long-term strategic vision for a prosperous , modern , competitive and climate neutral economy, Com(2018) 773 (2018) 114.

[20]   EUROPEAN COMMISSION, The European Green Deal, EC, Brussels (2019) 47–65.

[21]   NUCLEAR ENERGY AGENCY, ORGANIZATION FOR ECONOMIC COOPERATION AND DEVELOPMENT, The Costs of Decarbonisation: System Costs with High Shares of Nuclear and Renewables, NEA/OECD, Paris (2019).

[22]   RUSSEL, R., SCHLANDT, J., Loop flows: why is wind power from northern Germany putting east European grids under pressure?, Clean Energy Wire (Dec. 2015).

[23]   TIETJEN, J.S., SCHUSSLER, R.P., The myth of the German renewable energy 'miracle', T&D World (Oct. 2017).

[24]   INTERNATIONAL ATOMIC ENERGY AGENCY, Opportunities for Cogeneration with Nuclear Energy, IAEA Nuclear Energy Series No. NP-T-4.1, IAEA, Vienna (2017).

[25]   INTERNATIONAL ENERGY AGENCY, World Energy Outlook 2018, IEA, Paris (2018).

[26]   WORLD HEALTH ORGANIZATION, Burden of Disease from Household Air Pollution for 2012, WHO, Geneva (2014).

[27]   KHARECHA, P.A., HANSEN, J.E., Prevented mortality and greenhouse gas emissions from historical and projected nuclear power, Environ. Sci. Technol. **47** (2013) 4889.

[28]   ORGANIZATION FOR ECONOMIC COOPERATION AND DEVELOPMENT, The Security of Energy Supply and the Contribution of Nuclear Energy, Nucl. Energy **28** (2010) 171.

[29]   BRAGG-SITTON, S.M. et al., Nuclear–renewable Hybrid Energy Systems: 2016 Technology Development Program Plan, INL, Idaho Falls, ID (2016).

[30] GABBAR, H.A., ABDUSSAMI, M.R., Feasibility Analysis of Grid-Connected Nuclear–renewable Micro Hybrid Energy System, Proc. 7th Int. Conf. Smart Energy Grid Eng., SEGE 2019, IEEE (2019) 294–298.

[31] WORLD NUCLEAR ORGANIZATION, Nuclear Power Reactors, https://world-nuclear.org/information-library/nuclear-fuel-cycle/nuclear-power-reactors/nuclear-power-reactors.aspx

[32] WORLD NUCLEAR ASSOCIATION, Small Nuclear Power Reactors, https://www.world-nuclear.org/information-library/nuclear-fuel-cycle/nuclear-power-reactors/small-nuclear-power-reactors.aspx

[33] GABBAR, H.A., Smart Energy Grid Engineering, 1st edn, Elsevier, Amsterdam (2016).

[34] US DEPARTMENT OF ENERGY, Technology Readiness Assessment Guide, Springer, Berlin (2011).

[35] GOUGAR, H.D., BARI, R.A., KIM, T.K., SOWINSKI, T.E., WORRALL, A., Assessment of the Technical Maturity of Generation IV Concepts for Test or Demonstration Reactor Applications, INL, Idaho Falls, ID (2015).

[36] US DEPARTMENT OF ENERGY, 2014 Technology Readiness Assessment — Overview, DOE, Washington, D.C. (2015).

[37] CARMACK, W.J., BRAASE, L.A., WIGELAND, R.A., TODOSOW, M., Technology readiness levels for advanced nuclear fuels and materials development, Nucl. Eng. Des. **313** (2017) 177–184.

[38] OFFICE OF NUCLEAR ENERGY, 5 Fast Facts about Nuclear Energy, https://www.energy.gov/ne/articles/5-fast-facts-about-nuclear-energy

[39] WORLD NUCLEAR ASSOCIATION, The Inquiry into the Prerequisites for Nuclear Energy in Australia, WNA, London (2019).

[40] DENHOLM, P., KING, J.C., KUTCHER, C.F., WILSON, P.P.H., Decarbonizing the electric sector: Combining renewable and nuclear energy using thermal storage, Energy Policy **44** (2012) 301.

[41] RUTH, M., CUTLER, D., FLORES-ESPINO, F., STARK, G., JENKIN, T., The Economic Potential of Three Nuclear–renewable Hybrid Energy Systems Providing Thermal Energy to Industry, Task No. SA15.1008, NREL, Washington, D.C. (2016).

[42] CALISKAN, H., DINCER, I., HEPBASLI, A., Exergoeconomic, enviroeconomic and sustainability analyses of a novel air cooler, Energy Build. **55** (2012) 747.

[43] SUMAN, S., Hybrid nuclear–renewable energy systems: a review, J. Clean. Prod. **181** (2018) 166.

[44] BRAGG-SITTON, S.M., BOARDMAN, R., RUTH, M., ZINAMAN, O., FORSBERG, C., Rethinking the Future Grid: Integrated Nuclear Renewable Energy Systems: Preprint, 9th Nucl. Plants Curr. Issues Symp. Mov. Forw. (2014).

[45] CSIK, B.J., KUPITZ, J., Nuclear power applications: supplying heat for homes and industries, IAEA Bull. **39-2** (1997) 21.

[46] FOOD AND AGRICULTURE ORGANIZATION OF THE UNITED NATIONS, 'Energy-Smart' Food for People and Climate, FAO, Rome (2011).

[47] INTERNATIONAL RENEWABLE ENERGY AGENCY, Accelerating Geothermal Heat Adoption in the Agri-Food Sector — Key Lessons and Recommendations, IRENA, Abu Dhabi (2019).

[48] INTERNATIONAL ATOMIC ENERGY AGENCY, Opportunities for Cogeneration with Nuclear Energy, IAEA Nuclear Energy Series No. NP-T-4.1, IAEA, Vienna (2017).

[49] POPOV, D., BORISSOVA, A., Innovative configuration of a hybrid nuclear–solar tower power plant, Energy **125** (2017) 736.

[50] POPOV, D., BORISSOVA, A., Innovative configuration of a hybrid nuclear–parabolic trough solar power plant, Int. J. Sustain. Energy **37** (2018) 616.

[51] TUOMISTO, H., Nuclear District Heating Plans from Loviisa to Helsinki Metropolitan Area, Joint NEA/IAEA Expert Workshop on the Technical and Economic Assessment of Non-Electric Applications of Nuclear Energy, NEA/OECD, Paris (2013).

[52] OMMEN, T., MARKUSSEN, W.B., ELMEGAARD, B., Lowering district heating temperatures — impact to system performance in current and future Danish energy scenarios, Energy **94** (2016) 273.

[53] DE VALLADARES, M., Global Trends and Outlook for Hydrogen, IEA, Paris (2017).

[54] CONNELLY, E., ELGOWAINY, A., RUTH, M., Current Hydrogen Market Size: Domestic and Global, DOE, Washington, D.C. (2019).

[55] MIDREX, 2018 World Direct Reduction Statistics (2019), https://www.midrex.com/wp-content/uploads/Midrex_STATSbookprint_2018Final-1.pdf

[56] HYDROGEN COUNCIL, Hydrogen, Scaling Up (2017), https://hydrogencouncil.com/en/study-hydrogen-scaling-up/

[57] ANL, Light Duty Electric Drive Vehicles Monthly Sales Updates, https://www.anl.gov/esia/light-duty-electric-drive-vehicles-monthly-sales-updates

[58] SAMSUN, R.C., ANTONI, L., REX, M., STOLTON, D., Deployment Status of Fuel Cells in Road Transport: 2021 Update, Forschungszentrum Jülich, Jülich (2021).

[59] CALIFORNIA FUEL CELL REVOLUTION, The California Fuel Cell Revolution (2018), https://h2fcp.org/sites/default/files/CAFCR-Presentation-2030.pdf

[60] DEVLIN, P., MORELAND, G., DOE Hydrogen and Fuel Cells Program Record #18002 Industry Deployed Fuel Cell Powered Lift Trucks (2018), https://www.hydrogen.energy.gov/pdfs/18002_industry_deployed_fc_powered_lift_trucks.pdf

[61] INTERNATIONAL ATOMIC ENERGY AGENCY, Hydrogen Production Using Nuclear Energy, IAEA Nuclear Energy Series No. NP-T-4.2, IAEA, Vienna (2012).

[62] NAGAISHI, R., KUMAGAI, Y., "Radiolysis of water", Nuclear Hydrogen Production Handbook (YAN, X. L., HINO, R., Eds), CRC Press, Boca Raton (2011)

[63] INTERNATIONAL ATOMIC ENERGY AGENCY, Industrial Applications of Nuclear Energy, IAEA Nuclear Energy Series No. NP-T-4.3, IAEA, Vienna (2017).

[64] PERRET, R., Solar Thermochemical Hydrogen Production Research (STCH) Thermochemical Cycle Selection and Investment Priority, Sandia National Laboratories, Livermore, CA (2011).

[65] SUMMERS, W., BUCKNER, M., Hybrid Sulfur Thermochemical Process Development, DOE, Washington, D.C. (2006).

[66] RUSS, B., Sulfur Iodine Process Summary for the Hydrogen Technology Down-Selection, INL, Idaho Falls, ID (2009).

[67] US DEPARTMENT OF ENERGY, Hydrogen Production: Thermochemical Water Splitting, DOE, Washington, D.C. (2019).

[68] KASAHARA, S. et al., JAEA's R&D on the Thermochemical Hydrogen Production IS Process, HTR 2014 (2014).

[69] DEOKATTEY, S., BHANUMURTHY, K., VIJAYAN, P.K., DULERA, I. V, Hydrogen production using high temperature reactors: an overview, Adv. Energy Res. **1** (2013) 013–033.

[70] SCHMIDT, O. et al., Future cost and performance of water electrolysis: an expert elicitation study, Int. J. Hydrog. Energy **42** (2017) 30470.

[71] CARMO, M., FRITZ, D., MERGEL, J., STOLTEN, D., A comprehensive review on PEM water electrolysis, Int. J. Hydrog. Energy (2013).

[72] HOVSAPIAN, R., Role of Electrolyzers in Grid Services, DOE, Washington, D.C. (2017).

[73] HART, D., LEHNER, F., JONES, S., LEWIS, J., KLIPPENSTEIN, M., The Fuel Cell Industry Review 2018, E4tech (2018).

[74] US DEPARTMENT OF ENERGY, 3.1 Hydrogen Production, DOE, Washington, D.C. (2015).

[75] LEI, L., TAO, Z., WANG, X., LEMMON, J., CHEN, F., Intermediate-temperature solid oxide electrolysis cells with thin proton-conducting electrolyte and a robust air electrode, J. Mater. Chem. A **5** (2017) 22945–22951.

[76] TSIMIS, D., AGUILO-RULLAN, A., ATANASIU, M., ZAFEIRATOU, E., DIRMIKI, D., The status of SOFC and SOEC R&D in the European Fuel Cell and Hydrogen Joint Undertaking Programme, ECS Trans. **91** (2019) 9.

[77] TUCKER, M. et al., HydroGEN: High Temperature Electrolysis (HTE) Hydrogen Production, DOE, Washington, D.C. (2018).

[78] MILLER, A., DUFFEY, R., Integrating large-scale co-generation of hydrogen and electricity from wind and nuclear sources (NuWind©), IEEE EIC Climate Change Technology (2005).

[79] INTERNATIONAL ATOMIC ENERGY AGENCY, Nuclear–Renewable Hybrid Energy Systems for Decarbonized Energy Production and Cogeneration, IAEA-TECDOC-1885, IAEA, Vienna (2019).

[80] RUTH, M., CUTLER, D., FLORES-ESPINO, F., STARK, G., The Economic Potential of Nuclear–Renewable Hybrid Energy Systems Producing Hydrogen, NREL/TP-6A50-66764, JISEA, Golden, CO (2017).

[81] TAKEI, M., KOSUGIYAMA, S., MOURI, T., KATANISHI, S., KUNITOMI, K., Economical evaluation on gas turbine high temperature reactor 300 (GTHTR300), Nippon Genshiryoku Gakkai Wabun Ronbunshi, (2006).

[82] KIM, J.S., BOARDMAN, R.D., BRAGG-SITTON, S.M., Dynamic performance analysis of a high-temperature steam electrolysis plant integrated within nuclear–renewable hybrid energy systems, Appl. Energy **228** (2018) 2090.

[83] FRICK, K.L. et al., Evaluation of Hydrogen Production Feasibility for a Light Water Reactor in the Midwest, INL, Idaho Falls, ID (2019).

[84] UNDP UNITED NATIONS DEVELOPMENT PROGRAMME, Sustainable Development Goal 6 Targets (2018), https://sdgs.un.org/goals/goal6

[85] OFFICE OF ENERGY EFFICIENCY AND RENEWABLE ENERGY, Water Security Grand Challenge (2022), https://www.energy.gov/eere/water-security-grand-challenge

[86] HARES, S., The Cost of Clean Water: $150 Billion a Year, Says World Bank (2017),
https://www.reuters.com/article/us-global-water-health-idUSKCN1B812E

[87] INTERNATIONAL DESALINATION ASSOCIATION, Dynamic Growth for Desalination and Water Reuse in 2019,
https://idadesal.org/dynamic-growth-for-desalination-and-water-reuse-in-2019/

[88] INTERNATIONAL DESALINATION ASSOCIATION, The IDA Water Security Handbook 2018–2019, Global Water Intelligence, Media Analytics Ltd, Oxford (2019).

[89] SEPARATIONPROCESSES, Multi-Stage Flash Distillation, Separation Processes,
http://www.separationprocesses.com/Distillation/DT_Chp07a.htm

[90] SEPARATIONPROCESSES, Multi-Effect Distillation, Separation Processes,
http://www.separationprocesses.com/Distillation/DT_Chp07b.htm

[91] SEPARATIONPROCESSES, Vapour Compression (VC) Distillation, Separation Processes,
http://www.separationprocesses.com/Distillation/DT_Chp07c.htm

[92] AQUATECH, Desalination Plants: Six of the World's Largest, Aquatech, (Jan. 2019).

[93] AL-OTHMAN, A. et al., Nuclear desalination: a state-of-the-art review, Desalination (2019).

[94] NUSCALE, Meeting 21st Century Water Challenges,
https://www.nuscalepower.com/environment/clean-water

[95] YURMAN, D., X-Energy Signs on with Jordan for Four 75 MWe HTGR, Energy Central,
https://energycentral.com/c/ec/x-energy-signs-jordan-four-75-mwe-htgr

[96] CONCA, J., , How 1,500 Nuclear-Powered Water Desalination Plants Could Save The World From Desertification (2019),
https://www.forbes.com/sites/jamesconca/2019/07/14/megadroughts-and-desalination-another-pressing-need-for-nuclear-power/#71e8866a7fde .

[97] WORLD NUCLEAR ASSOCIATION, Nuclear Power in Saudi Arabia (2022),
https://www.world-nuclear.org/information-library/country-profiles/countries-o-s/saudi-arabia.aspx

[98] KIM, J.S., CHEN, J., GARCIA, H.E., Modeling, control, and dynamic performance analysis of a reverse osmosis desalination plant integrated within hybrid energy systems, Energy **112** (2016) 52.

[99] EPINEY, AARON S., et al., Case Study: Nuclear–Renewable–Water Integration in Arizona, INL, Idaho Falls, ID (2018).

[100] EPINEY, A. et al., Case Study: Integrated Nuclear-Driven Water Desalination-Providing Regional Potable Water in Arizona, INL, Idaho Falls, ID (2019).

[101] INTERNATIONAL ATOMIC ENERGY AGENCY, Successful Commissioning of Nuclear Desalination Plant in Pakistan, Nucl. Desalin. Newsl. (Sep. 2011) 4.

[102] SALEEM, M., Environmental impact assessment of nuclear desalination plant at KANUPP, Nucleus **47** (2010) 313–332.

[103] INTERNATIONAL ATOMIC ENERGY AGENCY, Desalination Economic Evaluation Program (DEEP), Computer Manual Series 14, IAEA, Vienna (2000).

[104] KHAN, S.U.-D., Karachi Nuclear Power Plant (KANUPP): As case study for techno-economic assessment of nuclear power coupled with water desalination, Energy **127** (2017) 372.

[105] BALOCH, M.H. et al., Hybrid energy sources status of Pakistan: an optimal technical proposal to solve the power crises issues, Energy Strateg. Rev. **24** (2019) 132.

[106] WORRELL, E., PRICE, L., MARTIN, N., HENDRIKS, C., MEIDA, L.O., Carbon dioxide emission from the global cement industry, Annu. Rev. Energy Environ. **26** (2001) 303.

[107] ALI, M.B., SAIDUR, R., HOSSAIN, M.S., A review on emission analysis in cement industries, Renew. Sustain. Energy Rev. **15** (2011) 2252.

[108] WANG, Q., LUO, J., ZHONG, Z., BORGNA, A., CO_2 capture by solid adsorbents and their applications: current status and new trends, Energy Environ. Sci. **4** (2011) 42.

[109] YU, C.H., HUANG, C.H., TAN, C.S., A review of CO_2 capture by absorption and adsorption, Aerosol Air Qual. Res. **12** (2012) 745.

[110] LEUNG, D.Y.C., CARAMANNA, G., MAROTO-VALER, M.M., An overview of current status of carbon dioxide capture and storage technologies, Renew. Sustain. Energy Rev. **39** (2014) 426.

[111] ATSONIOS, K., GRAMMELIS, P., ANTIOHOS, S.K., NIKOLOPOULOS, N., KAKARAS, E., Integration of calcium looping technology in existing cement plant for CO2 capture: process modeling and technical considerations, Fuel **153** (2015) 210.

[112] REICH, L., YUE, L., BADER, R., LIPIŃSKI, W., Towards solar thermochemical carbon dioxide capture via calcium oxide looping: a review, Aerosol Air Qual. Res. **14** (2014) 500.

[113] BENHELAL, E., ZAHEDI, G., SHAMSAEI, E., BAHADORI, A., Global strategies and potentials to curb CO_2 emissions in cement industry, J. Clean. Prod. **51** (2013) 142.

[114] FLAMANT, G., HERNANDEZ, D., BONET, C., TRAVERSE, J.P., Experimental aspects of the thermochemical conversion of solar energy; decarbonation of $CaCO_3$, Sol. Energy **24** (1980) 385.

[115] FLAMANT, G., GAUTHIER, D., BOUDHARI, C., FLITRIS, Y., A 50 kW fluidized bed high temperature solar receiver: heat transfer analysis, J. Sol. Energy Eng. **110** (1988) 313.

[116] LICHT, S. et al., STEP cement: solar thermal electrochemical production of CaO without CO_2 emission, Chem. Commun. **48** (2012) 6019.

[117] MEIER, A., BONALDI, E., CELLA, G.M., LIPINSKI, W., WUILLEMIN, D., Solar chemical reactor technology for industrial production of lime, Sol. Energy **80** (2006) 1355.

[118] MEIER, A. et al., Design and experimental investigation of a horizontal rotary reactor for the solar thermal production of lime, Energy **29** (2004) 811.

[119] MEIER, A., GREMAUD, N., STEINFELD, A., Economic evaluation of the industrial solar production of lime, Energy Convers. Manag. **46** (2005) 905.

[120] MEIER, A., BONALDI, E., CELLA, G.M., LIPINSKI, W., Multitube rotary kiln for the industrial solar production of lime, J. Sol. Energy Eng. **127** (2005) 386.

[121] SALMAN, O.A., KRAISHI, N., Thermal decomposition of limestone and gypsum by solar energy, Sol. Energy **41** (1988) 305.

[122] SCEATS, M.G., HORLEY, C.J., RICHARDSON, P., System and Method for the Calcination of Minerals, United States Patent No.: US8807993B2., 2014.

[123] HILLS, T.P., SCEATS, M., RENNIE, D., FENNELL, P., LEILAC, Low cost CO_2 capture for the cement and lime industries, Energy Procedia **114** (2017) 6166.

[124] HANEKLAUS, N., SCHRÖDERS, S., ZHENG, Y., ALLELEIN, H.-J., Economic evaluation of flameless phosphate rock calcination with concentrated solar power and high temperature reactors, Energy **140** (2017) 1148–1157.

[125] HANEKLAUS, N., ZHENG, Y., ALLELEIN, H., Stop smoking — tube-in-tube helical system for flameless calcination of minerals, Processes **5** (2017), 67.

[126] RUTH, M.F. et al., Nuclear–renewable hybrid energy systems: opportunities, interconnections and needs, Energy Convers. Manag. **78** (2014) 684.

[127] FORSBERG, C., Hybrid systems to address seasonal mismatches between electricity production and demand in nuclear renewable electrical grids, Energy Policy **62** (2013) 333.

[128] BURGALETA, J.I., ARIAS, S., RAMIREZ, D., Gemasolar: the first tower thermosolar commercial plant with molten salt storage system, SolarPACES Int. Conf. Sept. (2012) 11.

[129] DUNN, R.I., HEARPS, P.J., WRIGHT, M.N., Molten-salt power towers: newly commercial concentrating solar storage, Proc. IEEE **100** (2012) 504.

[130] ZHANG, Z. et al., Current status and technical description of Chinese 2 × 250 MWth HTR-PM demonstration plant, Nucl. Eng. Des. **239** (2009) 1212.

[131] ZHANG, Z., WU, Z., SUN, Y., LI, F., Design aspects of the Chinese modular high-temperature gas-cooled reactor HTR-PM, Nucl. Eng. Des. **236** (2006) 485.

[132] RUTH, M., CUTLER, D., FLORES-ESPINO, F., STARK, G., JENKIN, T., SIMPKINS, T., MACKNICK, J., The Economic Potential of Two Nuclear-Renewable Hybrid Energy Systems, Technical Report NREL/TP-6450-66073, USDOE, Washington, D.C. (2016).

[133] WOOD, R.A., BOARDMAN, R.D., PATTERSON, M.W., MILLS, P.M., Sensitivity of Hydrogen Production via Steam Methane Reforming to High Temperature Gas-Cooled Reactor Outlet Temperature Process Analysis, INL/TEV-962, INL, Idaho Falls, ID (2010).

[134] KUBIC, W.L., Nuclear-Power Ammonia Production, LA-UR-06-7180, LANL, Los Alamos, NM (2006).

[135] INL, T., Integration of High-Temperature Gas-Cooled Reactors into Industrial Process Applications, INL/EXT-11-23008, INL, Idaho Falls, ID (2009).

[136] INTERNATIONAL ATOMIC ENERGY AGENCY, Industrial Applications of Nuclear Energy, IAEA Nuclear Energy Series No. NP-T-4.3, IAEA, Vienna (2017).

[137] CHEN, Q. et al., Hybrid energy system for a coal-based chemical industry, Joule **2** (2018) 607–620.

[138] CHEN, Q.Q., TANG, Z.Y., LEI, Y., SUN, Y.H., JIANG, M.H., Feasibility analysis of nuclear–coal hybrid energy systems from the perspective of low-carbon development, Appl. Energy **158** (2015) 619–630.

[139] BREDIMAS, A., Results of a European industrial heat market analysis as a pre-requisite to evaluating the HTR market in Europe and elsewhere, Nucl. Eng. Des. **271** (2014) 41.

[140] BUCK, R., GIULIANO, S., Solar tower system temperature range optimization for reduced LCOE, AIP Conference Proceedings, Vol. 2126, AIP (2019) 030010.

[141] BREEZE, Pushing the Steam Cycle Boundaries, Power Engineering International, https://www.powerengineeringint.com/2012/04/01/pushing-the-steam-cycle-boundaries/

[142] SOLAR ENERGY TECHNOLOGIES OFFICE, Concentrating Solar Power Basics, DOE, https://www.energy.gov/eere/solar/articles/concentrating-solar-power-basics

[143] MURPHY, C. et al., The Potential Role of Concentrating Solar Power within the Context of DOE's 2030 Solar Cost Targets, NREL, Golden, CO (2019).

[144] INTERNATIONAL ATOMIC ENERGY AGENCY, Nuclear–Renewable Hybrid Energy Systems for Decarbonized Energy Production and Cogeneration, IAEA-TECDOC-1885, IAEA, Vienna (2019).

[145] BORISSOVA, A., Analysis and synthesis of a hybrid nuclear–solar power plant, BgNS Trans. **20** (2015) 58.

[146] QUOILIN, S., BROEK, M. VAN DEN, DECLAYE, S., DEWALLEF, P., LEMORT, V., Techno-economic survey of organic Rankine cycle (ORC) systems, Renew. Sustain. Energy Rev. **22** (2013) 168.

[147] KIM, N.J., NG, K.C., CHUN, W., Using the condenser effluent from a nuclear power plant for ocean thermal energy conversion (OTEC), Int. Commun. Heat Mass Transf. **36** (2009) 1008–1013.

[148] PARK, S.S., KIM, W. J., KIM, Y.H., HWANG, J.D., KIM, N.J., Regenerative OTEC systems using condenser effluents discharged from three nuclear power plants in South Korea, Int. J. Energy Res. **39** (2015) 397.

[149] ENERGY.GOV, How Microgrids Work (2022), https://www.energy.gov/oe/activities/technology-development/grid-modernization-and-smart-grid/role-microgrids-helping

[150] ENERGYSAGE, What Are Microgrids and How Do They Work? (2022), https://news.energysage.com/what-are-microgrids/.

[151] BOARDMAN, R.D., et al., Evaluation of Non-Electric Market Options for a Light-Water Reactor in the Midwest, INL, Idaho Falls, ID (2019).

[152] EPINEY, A.S., CHEN, J., RABITI, C., Status on the Development of a Modeling and Simulation Framework for the Economic Assessment of Nuclear Hybrid Energy, INL, Idaho Falls, ID (2016).

[153] INTERNATIONAL ATOMIC ENERGY AGENCY, Stakeholder Involvement Throughout the Life Cycle of Nuclear Facilities, IAEA Nuclear Energy Series No. NG-T-1.4, IAEA, Vienna (2011).

[154] US DEPARTMENT OF ENERGY, Consent–Based Siting Process for Consolidated Storage and Disposal Facilities for Spent Nuclear Fuel and High-Level Radioactive Waste (2017), https://www.energy.gov/sites/prod/files/2017/01/f34/Draft%20Consent-Based%20Siting%20Process%20and%20Siting%20Considerations.pdf

[155] NATIONAL ENERGY BOARD, Canada's Energy Future 2018: Energy Supply and Demand Projections to 2040, CEB, Calgary (2018).

[156] BENAHMED, F., WALTER, L., Clean energy targets are trending, Third Way, https://www.thirdway.org/graphic/clean-energy-targets-are-trending

[157] DENHOLM, P., O'CONNELL, M., BRINKMAN, G., JORGENSON, J., Overgeneration from Solar Energy in California. A Field Guide to the Duck Chart, NREL, Golden, CO (2015).

[158] LOKHOV, A., Load-following with nuclear power plants, NEA News (2011).

[159] NUCLEAR ENERGY AGENCY, Technical and Economic Aspects of Load Following with Nuclear Power Plants, Nuclear Development, NEA/OECD, Paris (2011).

[160] INTERNATIONAL ATOMIC ENERGY AGENCY, Non-baseload Operation in Nuclear Power Plants: Load Following and Frequency Control Modes of Flexible Operation, IAEA Nuclear Energy Series No. NP-T-3.23, IAEA, Vienna (2018).

[161] CANY, C. et al., Nuclear and intermittent renewables: two compatible supply options? The case of the French power mix, Energy Policy **95** (2016) 135–146.

[162] CHEN, Z., CAI, J., LIU, R., WANG, Y., Preliminary thermal hydraulic analysis of various accident tolerant fuels and claddings for control rod ejection accidents in LWRs, Nucl. Eng. Des. **331** (2018) 282.

[163] ABDELHAMEED, A.A.E., NGUYEN, X.H., LEE, J., KIM, Y., Feasibility of passive autonomous frequency control operation in a soluble-boron-free small PWR, Ann. Nucl. Energy **116** (2018) 319.

[164] ABDELHAMEED, A.A.E., LEE, J., KIM, Y., Physics conditions of passive autonomous frequency control operation in conventional large-size PWRs, Prog. Nucl. Energy **118** (2020) 103072.

[165] FORSBERG, C.W., Variable and assured peak electricity production from base-load light-water reactors with heat storage and auxiliary combustible fuels, Nucl. Technol. 205 (2019) 377–396.

[166] YU, H., HARTANTO, D., MOON, J., KIM, Y., A conceptual study of a supercritical CO_2-cooled micro modular reactor, Energies **2015** (2015) 13938–13952.

[167] TALBOT, P.W. et al., Analysis of Differential Financial Impacts of LWR Load-Following Operations, INL/EXT-19-55614-Rev001, USDOE, Idaho Falls, ID (2019).

[168] HÄRDLE, W.K., TRUECK, S., The dynamics of hourly electricity prices, discussion paper (2010), https://ssrn.com/abstract=2894267

[169] KHNP, Pumped-Storage Power Status, http://khnp.co.kr/eng/content/762/main.do?mnCd=EN040301.

Annex I.

LEGEND OF GRAPHICS

This annex contains the legend of graphics for the figures in the publication (see Fig. A–1).

 Nuclear Reactor

 Nuclear Power Plant

 Small Modular Reactors

 Renewables

 Solar Photovoltaic

Concentrated Solar Power

Wind Power

 Hydropower

 Biomass

 Geothermal

Heat Exchanger 1

Heat Exchanger 2

Turbine-Generator Set

 Heat Storage Tank

Electrical Storage / Battery

Heat

Electricity

Industry

 Residential

Electrical Grid

Greenhouses

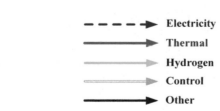
Electricity
Thermal
Hydrogen
Control
Other

FIG. A–1. Legend of graphics.

Abbreviations

ABWR	advanced boiling water reactor
APS	Arizona Public Service
CHP	combined heat and power
CCGT	combined cycle gas turbine
CSP	concentrated solar power
DERs	distributed energy resources
EJ	exajoule
FCEV	fuel cell electric vehicle
GHG	greenhouse gas
GTHTR	gas turbine high temperature reactor
GW(e)	gigawatt electrical
HES	hybrid energy system
HPS	high pressure steam
HTE	high temperature electrolysis
HTGR	high temperature gas reactor
HTF	heat transfer fluid
HTSE	high temperature steam electrolysis
ICL	intermediate coupling loop
IEA	International Energy Agency
IPS	intermediate pressure steam
KPIs	key performance indicators
LHV	lower heating value
LPS	low pressure steam
LTE	low temperature electrolysis
MED	multieffect distillation
MHES	microhybrid energy system
Mt	megatonnes
MSF	multistage flash
MW	Megawatt
NES	Nuclear Energy Series
NH_3	ammonia
NPP	nuclear power plant
OTEC	ocean thermal energy conversion
PEM	polymer electrolyte membrane
PVGS	Palo Verde Generating Station
R&D	research and development
RES	renewable energy source
RO	reverse osmosis
SOEC	solid oxide electrolysis cell
SMR	small modular reactor
TEA	technoeconomic assessment
TRL	technology readiness level
VC	vapour compression

CONTRIBUTORS TO DRAFTING AND REVIEW

Bragg-Sitton, S. M.	Idaho National Laboratory, USA
Caliskan, H.	Usak University, Türkiye
Debelak, K.	GEN Energija, Slovenia
Gaber, H.	University of Ontario Institute of Technology (UOIT), Canada
Haneklaus, N.	Rheinisch-Westfälische Technische Hochschule Aachen (RWTH Aachen University), Germany
Jevremovic, T	International Atomic Energy Agency
Arif, A.	Pakistan Atomic Energy Commission (PAEK), Pakistan
Kim, Y. H.	KAIST, Republic of Korea
Ruth, M.	National Renewable Energy Laboratory, USA
Tucek, K.	European Commission
Van Heek, A.	International Atomic Energy Agency

Technical Meeting

Vienna, Austria: 22–25 October 2018

Consultants Meetings

Vienna, Austria: 11–14 February 2019, 16–19 December 2019

Structure of the IAEA Nuclear Energy Series*

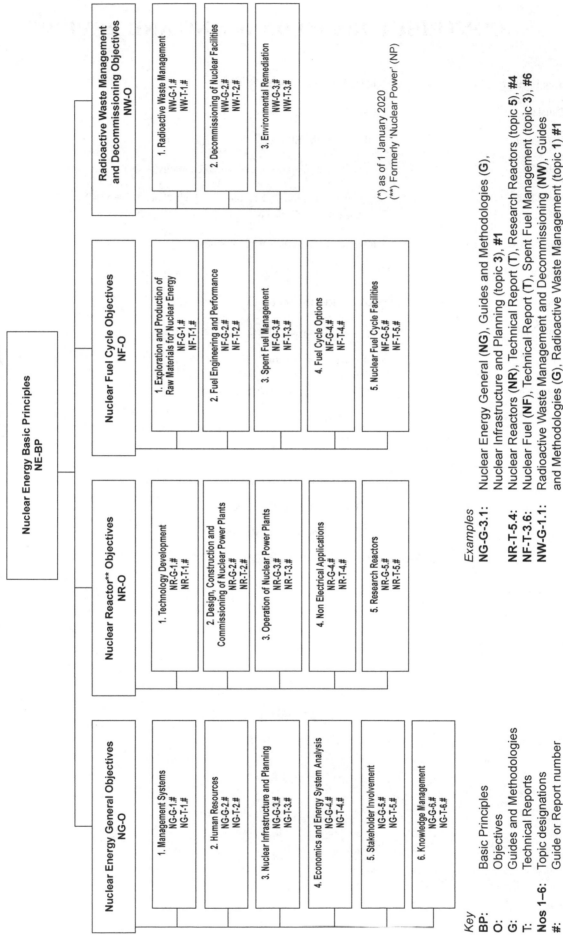

ORDERING LOCALLY

IAEA priced publications may be purchased from the sources listed below or from major local booksellers.

Orders for unpriced publications should be made directly to the IAEA. The contact details are given at the end of this list.

NORTH AMERICA

Bernan / Rowman & Littlefield

15250 NBN Way, Blue Ridge Summit, PA 17214, USA
Telephone: +1 800 462 6420 • Fax: +1 800 338 4550

Email: orders@rowman.com • Web site: www.rowman.com/bernan

REST OF WORLD

Please contact your preferred local supplier, or our lead distributor:

Eurospan Group

Gray's Inn House
127 Clerkenwell Road
London EC1R 5DB
United Kingdom

Trade orders and enquiries:

Telephone: +44 (0)176 760 4972 • Fax: +44 (0)176 760 1640
Email: eurospan@turpin-distribution.com

Individual orders:

www.eurospanbookstore.com/iaea

For further information:

Telephone: +44 (0)207 240 0856 • Fax: +44 (0)207 379 0609
Email: info@eurospangroup.com • Web site: www.eurospangroup.com

Orders for both priced and unpriced publications may be addressed directly to:

Marketing and Sales Unit
International Atomic Energy Agency
Vienna International Centre, PO Box 100, 1400 Vienna, Austria
Telephone: +43 1 2600 22529 or 22530 • Fax: +43 1 26007 22529
Email: sales.publications@iaea.org • Web site: www.iaea.org/publications